I0117228

THE WEATHER REPORT

THE WEATHER REPORT
A JOURNEY THROUGH UNSETTLED CLIMATES

Andrew Ross

Philadelphia, PA
Brooklyn, NY
commonnotions.org

The Weather Report: A Journey Through Unsettled Climates
© Andrew Ross

This edition © 2025 Common Notions
This is work is licensed under the Creative Commons Attribution-
NonCommercial 4.0 International. To view a copy of this license, visit https://
creativecommons.org/licenses/by-nc/4.0/.

ISBN: 978-1-945335-44-0 | eBook ISBN: 978-1-945335-58-7
Library of Congress Number: 2025941331

10 9 8 7 6 5 4 3 2 1

Common Notions   Common Notions
c/o Interference Archive   c/o Making Worlds Bookstore
314 7th St.   210 S. 45th St.
Brooklyn, NY 11215   Philadelphia, PA 19104

www.commonnotions.org
info@commonnotions.org

Discounted bulk quantities of our books are available for organizing, educational, or fundraising purposes. Please contact Common Notions at the
address above for more information.

Cover design by Josh MacPhee
Layout design and typesetting by Sydney Rainer
Printed by union labor in Canada on acid-free paper

CONTENTS

"Our greatest investment is Planet Earth," COP28 climate summit, Dubai, December 2023.

INTRODUCTION

To write *The Weather Report*, I tasked myself with a mission—to revisit and report from four locations where I had previously done research and written books: Palestine, the UAE, Arizona, and China. In each case, I had not been back for several years. I pledged to set out with no overarching thematic goals, but rather to respond to people and conditions on the ground as I found them. My hope was that the book—if it actually became one—would grow organically. It would reflect the sequential order of my visits to Ramallah, Dubai, Phoenix and Shanghai, and yield a storyline along the way. This plan fell apart much more quickly than I had bargained for. Time and again, a subterranean current of thought broke through to the surface of my mind, demanding my attention. Over the years, I had stored away many ideas for a future book about population and climate change. Although I had not foreseen it, this was the book that *The Weather Report* seemed to want to be.

Written between the summer of 2023 and fall 2024, this book begins and ends among Palestinians, whose deadly ordeal and steadfast resistance became an unavoidable focus of the whole world during that period of time. The genocide in Gaza attained a catastrophic dimension that mapped onto the environmental alarm signals. It was a year that saw records broken in every category of climate impacts—soaring temperatures, devastating floods, merciless droughts, sweeping wildfires, calamitous typhoons and hurricanes, and escalating rates of species extinction and ice sheet melt. Greenhouse gas emissions, the prime driver of these life-threatening events, also soared in 2023 to a record fifty-seven gigatons of CO_2 equivalent. Gaza's "second Nakba" ignited a global justice movement on a scale that resonated with the countless millions who have marched for climate justice. For people all over the world, standing up for Palestine was not simply an occasion for showing solidarity with a brutalized people; it spoke to our own hopes for liberation and a future not foreclosed by racial hatred, crushing debt, neocolonial militarism, corporate gluttony, and ecosystem collapse. That is one of more expansive meanings behind the movement slogan "Palestine is everywhere."

The gruesome spectacle of lives and lands devastated for generations to come is a reminder that ecocide has long been part of the colonial war waged against Palestinians. The fossil-fueled violence of the US and Israeli war machine took a ruinous toll on ecosystems already struggling with climate stress, while the denial of water, the most threatened resource in the Middle East, was deployed as a frontline weapon. However, it was the prospect of a mass eradication of the Gazan population that evoked the *existential* character of the war's outcome. Images of rubble-strewn cities and reports about the erasure of entire generations of families from Gaza's civil registry were apocalyptic in ways that we have come to associate with future scenarios of ecological crisis in disaster movies and science fiction.

From all accounts, it seemed the grisly future was arriving sooner than expected in most parts of the world, or at least earlier than the climate forecasters had predicted. In 1991, when *Strange Weather* (my first book about climate change) was published, it was still reasonable to talk about *mitigation* strategies for reducing anthropogenic emissions in order to minimize any crisis in the making.[1] The Inter-governmental Panel on Climate Change's (IPCC) First Assessment Report, issued the year before, suggested as much, hedged around with its share of qualifying language about "uncertainties." By 2024, the crisis was very much upon us, and the talk was all about *adaptation* measures to stave off, and essentially survive, only the worst impacts. A new type of climate melancholia, diagnosed by psychologists as a form of clinical depression, had taken hold among portions of the general population; manifest less in trauma from extreme weather events than by mourning for the future, or grief for lost species and ecosystems.[2]

In the past three decades, the movement for climate justice has gained strength as a corrective to the presumed authority of voices from the Global North to decide on climate policy. If the way forward is not to be a death march for large sectors of humanity, then the legacy of five hundred years of colonial plunder and extraction must be reckoned with and accounted for. The struggle to even out and democratize the environmental field of action has been the most important development over that period of time, and it has realigned the political energies of the ecologically minded. Advocates for climate justice have succeeded in shifting the focus away from an exclusively technocratic discussion of carbon budgets, greenhouse gas protocols, ice-albedo feedback, or the particular fetish of the Atlantic Meridional Overturning Circulation (i.e., the Gulf Stream). Climate change is now more of a civil rights cause, deeply underpinned by accountability for the debts owed to the most vulnerable communities who are also the least responsible for the problem.

---

1   Andrew Ross, *Strange Weather: Culture, Science and Technology in the Age of Limits* (London: Verso, 1991).

2   Britt Wray, *Generation Dread: Finding Purpose in an Age of Climate Crisis* (New York: Knopf, 2022); Ashlee Cunsolo and Karen Landman, eds., *Mourning Nature: Hope at the Heart of Ecological Loss and Grief* (Montreal: McGill-Queen's University Press, 2020); Shawna Weaver, *Climate Grief: From Coping to Resilience and Action* (Seattle: Lantern, 2023); Paolo Cianconi et al., "Eco-Emotions and Psychoterratic Syndromes: Reshaping Mental Health Assessment Under Climate Change," *Yale Journal of Biology and Medicine* 96, no. 2 (June 2023): 211–26.

Like all popular left causes, the struggle for climate justice proceeds along an unstable path, identified by Mike Davis as the familiar Gramscian oscillation between "pessimism of the will" and "optimism of the imagination."[3] Nothing we have learned about the tireless determination of fossil capitalists can lead us to expect a moderation of their will to extract every last barrel of oil and sell most of it to poor countries. After a brief interregnum (2019–2022) of climate commitments on the part of oil companies, investment banks, and fossil producer states, they all doubled down on the business of financing and building new extraction facilities. Yet there is much hope to be drawn from the dignity and momentum of popular ecological resistance—often labeled the "environmentalism of the poor"—acting in conjunction with the broad spectrum of anticapitalist movements. The dual spirit of that loose but fierce alliance animates these pages.

BOGUS SCARCITY

The belief that "there is not enough to go around," is fed by a river of mental dread that runs so deep it almost qualifies as primal. For most of us, this uneasy sentiment persists, despite all the evidence that human ingenuity has either freed us from absolute scarcity or is capable of doing so if only we could ensure a fair share of the commonwealth to everyone. The fear of not having enough because there are too many of us has taken on newly morbid dimensions in the face of the climate emergency. When the gates close around the well-to-do and preferred members of their servant class, will we be among those who are left outside to bake, drown, or starve? We have all seen such movies (my personal favorite is the spectacularly cheesy *2012*), or perused the books shelved under "cli-fi."

Thomas Malthus' conjectures cast a long shadow over how we think about the collectivity of our fellow humans. In his seminal 1798 essay on the relationship between population and food supply, he mainstreamed the idea that unchecked growth of the former would always outpace efforts to feed the new additions—and that chronic

---

3   Mike Davis, "Who Will Build the Ark?," *New Left Review* 61 (January–February 2010), https://newleftreview.org.

scarcity was the inescapable lot of the vast majority. Although he himself did not invoke the threat of overpopulation, "Malthusian" became a byword for those who fear-mongered about the topic. In fact, Malthus' primary goal in writing his essay was to offer a firm rebuke to influential advocates of social improvement, like William Godwin and Jean-Jacques Rousseau. For Malthus, their "optimistic" efforts to lift up the masses were a waste of time and resources. In his view, the poor would simply respond by reproducing more, thus adding to the burden of their existence.

A bleak version of this latter proposition would inspire successive waves of neo-Malthusians who singled out for condemnation excessive breeding on the part of "social undesirables" such as impoverished people and ethnic minorities.[4] These included the Social Darwinists (Herbert Spencer, William Graham Sumner, Francis Galton) in the 1880s and 1890s, who championed the fittest individuals over the "natural" losers; the eugenicists (Fairfield Osborn, Lothrop Stoddard, Madison Grant) in the 1920s and 1930s, who wanted to preserve the superior races by engineering the gene pool; and the ecologists (William Vogt, Paul and Anne Ehrlich, Garret Hardin, John Holdren, Lester Brown) in the 1950s and 1960s, who identified population growth in the Global South as the cause of impending mass starvation, if not outright environmental collapse.[5]

Despite plummeting fertility rates in most parts of the world, neo-Malthusianism has not run its course; it has only assumed new forms. In liberal circles, overpopulationists have tempered their zeal, having absorbed the feminist critiques of coercive birth control programs that were imposed on developing countries from the 1960s onwards. Today, their demographic pathway toward climate stabilization is guided by respect for reproductive justice, women's

---

4 Allan Chase, *The Legacy of Malthus: The Social Costs of the New Scientific Racism* (New York: Knopf, 1977).
5 Thomas Robertson, *The Malthusian Moment: Global Population Growth and the Birth of American Environmentalism* (New Brunswick, NJ: Rutgers University Press, 2011); Betsy Hartman, *Reproductive Rights and Wrongs: The Global Politics of Population Control* (Boston: South End Press, 1987); Ian Angus and Simon Butler, *Too Many People? Population, Immigration, and the Environmental Crisis* (Chicago: Haymarket, 2011).

empowerment, and LGBTQ+ rights. But the much darker side of the Malthusian legacy has also returned. Population shrinkage in affluent countries has unleashed a fresh wave of resentment about the loss of demographic majorities on the part of traditionally dominant, white ethnic groups. The far-right resolve to stop "the great replacement" of white people has gained ground through spreading disinformation about climate-induced scarcity. Surely, the white nationalists warn us, there won't be enough to go around if the borders are not sealed. These scare tactics have helped fuel the rise of eco-fascism. What Betsy Hartmann called the "greening of hate" has revolved around the defense of native "blood and soil" from an inroad of immigrants whose foreign ways put undue pressure on scarce resources.[6] This imagined invasion force includes climate refugees, who are now on the move in significant numbers.

Malthus reasoned that famine acts as a population control on communities (China was his prime example) who cannot exercise self-restraint over their reproductive activity. This was one of his many errors, including the baseless proposition that population increases geometrically while food supply only increases arithmetically. The main causes of famine, as Amartya Sen and others have long established, are maldistribution—there usually *is* enough to go around—and the market commodification of food that propels prices beyond what people can afford.[7] Climate impacts and warfare are also frequent factors, but overpopulation is not. The famine in Gaza had nothing to do with the fact that it is one of the most densely populated places on the planet; it is the result of one group of people intentionally starving another. Notwithstanding extreme weather, infestations, or plant diseases, human-made scarcity is usually the

---

6   Betsy Hartmann, "The Greening of Hate: An Environmentalist's Essay," *Greenwash: Nativists, Environmentalism and the Hypocrisy of Hate* (July 2010): 13–15; Andreas Malm and the Zetkin Collective, *White Skin, Black Fuel: On the Danger of Fossil Fascism* (London: Verso, 2022); John Hultgren, *Borders Walls Gone Green: Nature and Anti-Immigrant Politics in America* (Minneapolis: University of Minnesota Press, 2015); Ghassan Hage, *Is Racism an Environmental Threat?* (Cambridge: Polity, 2017).

7   Amartya Sen, *Poverty and Famines: An Essay on Entitlement and Deprivation* (Oxford: Oxford University Press, 1981).

reason for an extreme food shortage. Want was once the scourge of humanity, but that was before the manufacture and maintenance of scarcity became a central dynamic of capitalist markets. A non-capitalist alternative, as Murray Bookchin argued, would have to be a *post-scarcity society*, where people are free of such artificial constraints and vestigial fears about there not being enough to go around.[8]

As for climate change impacts, their root cause is anthropogenic, but the conditions of scarcity they give rise to are not directly planned or manipulated, as, say, a cartel might engineer a shortage of avocados or crude oil for profit. A parching drought might drive farmers off their land, a flood may wash away households on low-lying land, and a hurricane may devastate an entire island, but unless some weather modification scheme has played a role, it is difficult to show that specific individuals or groups have *intentionally* directed these tragic outcomes. Yet we can say fossil producers and high carbon emitters are ultimately responsible for these events if climate change has amplified the destruction. Indeed, the scientific evidence of causal agency is clear enough to warrant lawsuits against oil companies or to make claims on the climate debts owed by carbon-rich countries at the International Court of Justice (ICJ), where one hundred nations, led by the vulnerable Pacific Island states, brought a landmark climate justice case in December 2024.[9] Climate-driven harm is a new kind of injustice in our world, requiring new kinds of moral adjudication and legal remedy. All the more reason, as I learned from my field visits, for the need to distinguish it from the hardship caused by artificially manufactured scarcity or from the stinginess of nature.

In direct lineage from Malthus himself, the specter of food shortage has long consumed the attention of overpopulationists. For example, ecologist Lester Brown's widely read book *Who Will Feed China?* (1995) highlighted this fixation by focusing on China's

---

8 Murray Bookchin, *Post-Scarcity Anarchism* (San Francisco: Ramparts Press, 1971).
9 The ICJ case (*Obligations of States in Respect of Climate Change*, https://www. icj-cij.org/case/187) followed decisions from two other international courts—the International Tribunal on the Law of the Sea and the European Court of Human Rights—determining that states have obligations in relation to climate change, both towards other states and towards their citizens.

mounting population and shrinking cropland. His title also suggests a postcolonial version of the "white man's burden," i.e., it is up to Westerners to oversee the development of countries that will not shoulder their own responsibilities. For reasons of capacity, I have assumed here that world hunger, while acute, is primarily a problem of maldistribution and market dynamics, and so, with the exception of my report on agro-ecological initiatives among West Bank farmers, *The Weather Report* steers away from food toward water and energy. For obvious reasons, energy lies at the heart of the climate crisis. Globally, water insecurity is by far the most prevalent form of scarcity. Fresh water is more difficult to (re)distribute than food or energy; there is no more of it today than there was at the dawn of the Holocene and as a result of climate change increasingly there is less. According to the UN, more than half of the world's population is water insecure for at least parts of the year, and one quarter is severely water-stressed. One in four people on the planet (or 2.2 billion) lack safe drinking water, while almost half the world's population lacks access to safely managed sanitation.[10]

Global estimates like these are revealing, but the political character of such resource challenges varies quite sharply from place to place, as the four "unsettled climates" of this book demonstrate. The social footprint of water and energy in Israel/Palestine has little in common with their stamp on life in the UAE, even though both countries are water-stressed and boast an energy surplus. Many other differences presented themselves in the course of my visits. For example, population anxiety in China does not assume anything like the kind of racial dimension that it does in Arizona. So, too, the libertarian political culture of Phoenix has no correlate in Shanghai or Dubai, while the settler-colonial legacy of the US Southwest bears only a passing resemblance to the Zionist counterpart of land acquisition by Israel in Palestine. Engaging with these and other differences is important to how we think globally about climate justice but are moved to act locally, in response to conditions that may not apply elsewhere.

---

10 Richard Connor, "The United Nations World Water Development Report 2024: Water for Prosperity and Peace," *UNESCO World Water Assessment Programme*, March 19, 2024, https://www.un.org.

Advocates who reduce all climate action to numerical targets—stay below 1.5°C of global heating or 450 parts per million (ppm) of atmospheric carbon (422 ppm at the time of writing)—have little time for these regional variations, or for the maps of uneven risk and harm that matter most to climate justice activists. Their aim is to hit the arithmetical goals by any means necessary, even as the target-chasing—"keep 1.5°C alive"—is more and more illusory. As a result, their carbon calculations tend to skate over the history and currency of power relations and social stratification, along with most of what makes us capable, as humans, of rising to the task of meeting these targets: cultural custom, social reason, intercommunal fairness, and mutual care.[11] Above all, they cannot tell us how to survive and thrive, as we surely will, in a world irrevocably altered by climate change. In that spirit, The Weather Report tries to provide a more holistic version of the challenges I found in the four locations I visited, while highlighting a local justice initiative in each to show why there is reason to continue to practice hope.

FOUR CLIMES

The book begins in August 2023 among West Bank villagers, with my interviews with them about the water shortages they experience at the hands of Israeli authorities who manipulate and wield hydrological flows as an instrument of eco-apartheid. Parts of the West Bank see an annual rainfall greater than London, and sit on top of a vast aquifer, yet residents are lucky to get piped water in their houses once a week during the hot summer season. In rural areas, the denial and destruction of water supply is pushing dwellers off their lands. The backdrop for this "colonization through dehydration" is the history of Zionist settler anxiety about demographic supremacy in the lands of historic Palestine. Settlement in numbers sufficient enough to yield a Jewish majority has always been dependent on strategies to secure and monopolize water supplies, beginning with the diversion

---

11 See Michelle Murphy, *The Economization of Life* (Durham, NC: Duke University Press, 2017). Wim Carton and Andreas Malm describe in detail the discursive war over efforts to meet the 1.5°C target in *Overshoot: How the World Surrendered to Climate Breakdown* (London: Verso, 2024), 26–54.

of the River Jordan in the 1940s and the creation of the National Water Carrier distribution network that extends today to service settlements that lie deep in the West Bank. The weaponization of war was taken to new heights in the assault on Gaza with its precision bombing of pipes, sewage facilities, and desalination plants. As the atrocities mounted, it was difficult to avoid concluding that the longstanding demographic obsession to outnumber Palestinians had moved beyond ethnic cleansing into attempted depopulation through genocidal means. On a return visit to the West Bank the following summer, I spent time with members of the new wave of agricultural cooperatives, whose efforts to do "resistance farming" through land and water reclamation present a new kind of liberation front for Palestinian climate justice.

In contrast to Israel/Palestine, the ruling class in the UAE is a small native-born minority whose privileges are secured by a vast servant pool of foreign workers and a treasury of petroleum wealth. My attendance at the COP28 climate summit in Dubai in December 2023 allowed me to report from the cockpit of Gulf power, where climate diplomacy was being cynically managed in one of the world's least sustainable cities. It was clear that energy policy was on a dual track among the COP's leading fossil producers—net-zero plans at home, and stepped-up hydrocarbon exports to countries that cannot afford the investments in green technologies. I also found that the political freedoms and labor rights of migrants in the Gulf's autocracies are just as key to the future of climate justice as the technological advances in clean energy that are lavishly showcased at COP summits. Heat-stricken workers recruited from climate-distressed regions are the new canaries in the coal mine, and so the struggle for their well-being could not be more indicative of a "just transition" to a low-carbon society.

Like the Middle East, Arizona is warming up and drying out faster than anywhere in its hemisphere. In an election year, immigration was guaranteed to be a hot-button issue, and so the border wars in this swing state were feeding off the longstanding anxiety about water-dependent population growth in a region increasingly beset by aridification. My reports from water-stressed Phoenix and the overpoliced border region lay bare the deadly political and economic consequences of this volatile combination. On my visits to the

US-Mexico border zone, I accompanied humanitarian volunteers dispensing aid to border crossers who were coming from all over the world and whose plight was described by one local activist as "a thermometer that measures the temperature of what's happening in their countries of origin." Their caring ethos of welcome and assistance stands alongside the practices of Arizona's Indigenous "water protectors" as models of resilience and counterpower in a region where the harsh "lifeboat calculus" of population triage is now being administered by the Trump administration's increasingly militarized border security forces.

The Weather Report ends with my return to Shanghai in the grip of a "demographic winter." Accompanying my daughter to her provincial orphanage seventeen years after her adoption is part of a story I tell about the legacy of the state's one-child policy. With fertility rates in free fall, China's ruling party has GDP targets to maintain on behalf of the world's investors. The consequence of tying population control to economic development during the reform era has borne bitter fruit in unforeseen ways. In my interviews with Shanghainese women who are proving nonresponsive to state pressure to have children, I explored the contradictions of sustainable growth-driven development through the lens of social reproduction. China's command economy has allowed it to take the undisputed global lead in clean energy production, underpinned by the Communist Party's new governing template of "ecological civilization," but this large-scale initiative has come at the cost of popular participation and is only occurring within the framework of GDP growth. My Shanghai interviews foreshadow an eco-future in which women's empowerment guides the adaptation of state socialism to a new model of community care.

THE CODA

In the conclusion to this book I grapple with the need to find respite from the apocalyptic funk that permeates so many of our responses to the climate emergency. As someone who is allergic to despair, I tried, in my field trips, to look out for examples of effective or productive action. Without such encounters, it is easy to succumb to fatalism or defeatism, especially when faced with the ominous "greenlash" of a capitalist economy briskly recommitting to the expansion of fossil extraction with full-throated backing from the likes of Donald Trump. As Rebecca Solnit puts it, "I respect despair as an emotion but not as an analysis." To deflect the corrosive influence of "climate doomers," she insists that it is acceptable, and maybe even necessary, "to be hopeful and heartbroken at the same time."[12] Or, in Mariame Kaba's often-cited words, to accept that "hope is a form of discipline . . . we have to practice it every single day."[13]

Part of the work of curbing climate doom is to hold onto a vision of the salvaged future, even under the grimmest circumstances.[14] It is undeniable that we will be living in a quite altered biophysical world in the decades to come; the shape of that landscape is not settled though parts of it are already foretold. How we live on that unstable planet is still an open political question. We know how the worst scenarios are likely to play out; a starker version of eco-apartheid than is currently the case. The alternatives are not difficult to imagine or even agree on. A desirable green future would be a world of public affluence and community empowerment, where work is gratifying, energy is clean, technology is redemptive, housing and healthcare are more than adequate, and the supply of food and water are safe and sustainable. It would be a world where we commit to living within the limits of our ecosystems and to building a culture of interspecies solidarity. Other, more utopian, features of a just society that reach beyond profit and property are also necessary to dream about, to keep on talking about, and pushing for. Now more than ever.

Recently, proponents of the Green New Deal and advocates of degrowth have offered different pathways to that future. The former sees state power and public investment in renewable energy as the key to transitioning toward a green economy while maintaining standards of living. The latter envisages a more radical restructuring of our consumer society, with self-management of services and production

---

12 Rebecca Solnit, "We Can't Afford to be Climate Doomers," *The Guardian*, July 26, 2023, https://www.theguardian.com.
13 Mariame Kaba, *We Do This 'Til We Free Us: Abolitionist Organizing and Transforming Justice* (Chicago: Haymarket, 2021).
14 Out of the Woods Collective, *Hope against Hope: Writings on Ecological Crisis* (New York: Common Notions, 2020).

at the common center. But we may end up needing a hybrid economy which embraces both, where we do not have to choose as a matter of political principle and practice between "public" and "commons" as macro-level programs for distributing resources. Far from being mutually exclusive, one can be preferred over the other, according to context. Or we can embrace a public-commons model where officials who control provision of services agree to work alongside grassroots organizations in order to ensure community self-management. A decisive, and timely, shift to mass forms of clean energy, for example, is unlikely without the exercise of public power but democratic participation in decisions about generation, distribution, and administration needs to be locked in.[15] The provision of other goods—food, shelter, health care, clothing, culture, credit, information and learning—would certainly benefit from state backing and protection, but might also be better shared through networks of mutual aid. Our challenge, in the fickle Anthropocene, is to build a post-scarcity society that respects both modes of living and governing.

---

15 On the role of the state and public power, see Olúfẹ́mi O. Táíwò, "Giant Steps," and the range of responses in "What Is the State For?" a special issue of the *Boston Review* (February 2024).

Sabil fountain in the Old City, East Jerusalem, August 2024.

1

PALESTINE
THE LONG FRONT OF DEPOPULATION

"This summer, we are lucky if we get water in my home once in twenty days," reported Ramzy, a villager from nearby Jifna, as he filled his grimy four-gallon tanks from a tap near the open road. My translator colleague and I were standing at a spring just a few miles north of Ramallah/Al-Bireh, an urban area with an annual rainfall greater than London, but which, like the rest of Palestine, suffers from the imposition of artificial water scarcity at the hands of Israel.[1] "There is a great lake of water beneath us," Ramzy pointed out (referring to the bountiful Mountain Aquifer), "but we do not see any of its benefits. If we try to dig a well, we will be fined, and maybe worse." Like the others lined up with their containers, he was relying on the largesse of a man who had found a spring while digging the founda-

---

1 "Not Enough Water in the West Bank," *Visualizing Palestine*, https://101. visualizingpalestine.org.

tions for a new house and decided to make the water available to the public. I would later learn from the regional water service provider that the spring's water was not all that safe to drink from: it had been contaminated by the cesspits in the surrounding villages. But for Ramzy and his family, there were few alternatives. Water from the private tanker trucks that are ubiquitous on the West Bank's streets and roads comes at a steep price, and its quality is often not much better. The choice to line up at the spring was one of the many compromises he and others were forced to make under the Occupation; in some ways, it was the most existential.

During the summer of 2023, I returned to Palestine as its ongoing water crisis reached dangerous new heights. Next to the surge in settler aggression, anxiety about the lack of domestic water supply was the most common topic of discussion. And for many water-deprived households like Ramzy's, the quality of what they could obtain to drink was often not a priority. Among the factors contributing to the acute shortage were the unprecedented summer temperatures (the hottest to date); Israel's cruel reduction, by 25 percent, of water supply to the governorates of Hebron and Bethlehem; pressure on resources due to the post-COVID-19 influx of summer residents from the Palestinian diaspora; and the seizure of artesian springs by settlers all across the West Bank.

The daily supply to Israelis and to Jewish settlers is five to seven times greater than to the average Palestinian household, whose average consumption is almost 30 percent below the *minimum* amount of 100 liters per capita recommended by the World Health Organization (WHO).[2] Since the West Bank settlements are almost all connected to Israel's national water network, they have access to generously subsidized supplies. Settlers can always fill their swimming pools and irrigate their organic vineyards, even at the height of the region's scorching summers.

Under the 1995 Oslo Accords, Israel is entitled to siphon off a full 80 percent of the West Bank's aquifer groundwater. The remainder, allocated to the Palestinian Water Authority (PWA), has not

---

2 Amnesty International, "The Occupation of Water," November 29, 2017, https://www.amnesty.org.

come close to accommodating the sharp growth in population in the Occupied Territories, which today is home to 75 percent more residents than at the time of the Oslo agreements. Mekorot, Israel's parastate water company, controls almost all the wells, and its monopoly over supply ensures that water delivery need not even meet the agreed-upon quotas. In common with other essentials—electricity, work permits, tax revenue—Israeli authorities apply water restrictions as a form of collective punishment; during the First Intifada, and, in response to subsequent civil uprisings, communities were cut off from water, electricity, and fuel. Not unpredictably, this stark power to turn on and off the tap is also being used to accelerate the rate of Jewish settlement in the West Bank. Without steady provision of water to their villages or their farms, rural Palestinians are being pushed off their land and into the overbuilt and overcrowded cities. The goal is to maximize Israel's appropriation of territory, while further ghettoizing Palestinians in fewer and fewer population pockets. The latest land grabs are proceeding through the proactive *dehydration* of its occupants.

Awareness of the stark inequality in water provision is hardly new. Indeed, commentators have talked about "water apartheid" for some time now. The phrase appears often in the reports of human rights groups,[3] in the mainstream media,[4] and even some liberal Israeli critics have adopted the language.[5] But while apartheid talk has generated much-needed attention to Israel's injustices, it is an insufficient description of the colonial conditions in Palestine. In the public mind, "apartheid" suggests the maintenance of rule through a racial hierarchy upheld by Israeli courts; the strongest laws apply to seven million Israeli Jews, while lesser legal protections are afforded to the

---

3   Elisabeth Koek et al., "Water for One People Only: Discriminatory Access and 'Water-Apartheid' in the OPT," *Al-Haq* (2013), https://www.alhaq.org.

4   Mersiha Gadzo, "How Israel Engages in 'Water Apartheid,'" *Al Jazeera*, October 21, 2017, https://www.aljazeera.com.

5   For example, B'Tselem, Israel's premier information center on rights violations, notes that the water crisis "is an intentional outcome of Israel's deliberately discriminatory policy, which views water as another means for controlling the Palestinian population." Eyal Hareuveni, "Parched: Israel's Policy of Water Deprivation in the West Bank," *B'Tselem* (April 2023), https://www.btselem.org.

six separate classes of Palestinians; the two million citizens inside
Israel, three million occupied subjects in the West Bank, two million
Gazans confined to their open air prison, 360,000 permanent resi-
dents of East Jerusalem, 3,600 administrative detainees held without
trial in Israel's official prisons, and 2.4 million registered refugees in
Jordan, Lebanon, and Syria.

Yet the Occupation's daily business of displacement, depopulation, and land grabbing proceeds at a speed and on a scale that far
outpaces legal adjudication and innovation. Emboldened by the current far-right government coalition, whose leading members openly
advocate for the expulsion of Palestinians, settlers have been aided
and abetted by the Netanyahu administration's soldiers and administrators to obtain territory across the West Bank without regard for
the Israeli laws meant to deter them, not to mention the international
laws that are violated daily by the presence of 500,000 settlers in
everyday residence (and 230,000 more in East Jerusalem). In many
locations, as my Palestinian colleagues and I discovered in our travels,
the unofficial map of conquest is changing very rapidly. A road that
was safe the week before is now a no-go area. "You will be shot if you
go down there," we were told on more than one occasion.

This region was water-stressed long before global warming
picked up pace. Many of the villagers we met insisted that the next
war would be fought over water. Some commentators have even
argued that the 1967 Six-Day War (the *Naksa* for Palestinians) was
triggered by a stand-off with the Arab League over abstractions from
the Upper Jordan River basin. Aside from the longstanding state-level
conflicts over that watershed, clashes over local water resources have
become a daily occurrence in villages across the West Bank.

The military manipulation of water supply escalated just six
weeks after our 2023 field tours, in the aftermath of the Hamas attack
(Operation Al-Aqsa Flood, or *Toufan al-Aqsa*) on October 7. Israel's
immediate response indicated that Gaza's water supply would be cut
off in pursuit of military goals that went far beyond the usual template
of collective punishment. Two days after the attack, Defense Minister
Yoav Gallant declared: "I have ordered a complete siege on the Gaza
Strip. There will be no electricity, no food, no fuel, everything is
closed," adding that "we are fighting human animals and we are acting

accordingly."[6] The following day, Major General Ghassan Alian, the head of Israel's Coordination of Government Activities in the Territories (COGAT) turned up the rhetoric: "Human beasts are dealt with accordingly. Israel has imposed a total blockade on Gaza. No electricity, no water, just damage. You wanted hell, you will get hell."[7] This kind of dehumanizing language would later bolster the international legal case that Israel was committing genocide in Gaza—the worst crime against humanity. The military authorities carried through on the threats; its tactics of water warfare included not only the denial of Israeli-controlled water to civilian populations but also the bombardment of hydrological facilities—wells, wastewater facilities, and desalination plants—that were under Palestinian control in Gaza. By November 2024, the official estimate of deaths had risen to 44,000, but a July study published in *The Lancet* pointed out that this number did not account for tens of thousands buried under rubble and the number of "indirect deaths" due to destruction of health facilities, food and water distribution systems, and other public infrastructure. The authors estimated that the true death toll was closer to 186,000.[8] Another study in October 2024, based on physicians' observations, estimated 62,413 additional deaths from starvation, and another 5,000 from lack of access to care for chronic diseases.[9] In addition, as much as 80 percent of the Strip had been destroyed, and almost all of the population was displaced. No recorded war in history has come close to this level of intentional devastation.

By then, we were no longer talking about apartheid; the accusation of genocide had become firmly established in the minds and

---

6   Emanual Fabian, "Defense Minister Announces 'Complete Siege' of Gaza: No Power, Food or Fuel," *The Times of Israel*, October 9, 2023, https://www.timesofisrael.com.

7   "UN Human Rights Briefing on IHL," Office of the United Nations High Commissioner for Human Rights, June 19, 2024, https://www.unognewsroom.org.

8   "Gaza Toll Could Exceed 186,000, *Lancet* Study Says," *Al Jazeera*, July 8, 2024, https://www.aljazeera.com.

9   Sophia Stamatopoulou-Robbins, "The Human Toll: Indirect Deaths from War in Gaza and the West Bank, October 2023 Forward," *Costs of War*, Watson Institute, Brown University (October 7, 2024), https://watson.brown.edu/costsofwar/.

the lexicon of much of the world's appalled onlookers. A mountain of evidence for this charge was presented in the case brought by South Africa at the International Court of Justice, though the damning judgment passed down by the court had little impact on Israel's military conduct nor on the full-throated consent for it by Western powers. The only question that remained was whether the genocide had occurred as a discrete, if extreme, act of war, or whether it was the logical culmination of longstanding Israeli policies and practices, dating to the mass ethnic cleansing of the *Nakba* in 1948, and the Zionist dreams of conquest that drove it.

"Eliminating the native" is often cited as a key element of settler colonialism.[10] Taking the vagaries of history into consideration, it is difficult to dispute that the long-term tendency of Zionist settlement has been to get rid of the Indigenous population, whether by transfer, eviction, displacement, (self) deportation, or starvation. Over the years, liberal Zionists who pay lip service to coexistence with Palestinians have won temporary stays of execution, but they always lose their leverage to the maximalists in the face of a drummed-up threat. The horrific onslaught on Gazans brought to the surface of public attention a flood of expressed fantasies, on the part of Israeli officials and civilians alike, about exterminating the population, largely made up of Nakba refugees, or driving them out, or, at the very least, destroying their means of subsistence and thwarting their ability to reproduce for decades to come. These fantasies quickened after Donald Trump vowed, in February 2025, to clear the Gaza Strip of its population to make way for real estate development he described as a "Riviera of the Middle East."

Due to Israel's blockade of basic resources and the intermittent bombardment of the Strip known as "mowing the lawn," Gazans had been kept in a state of extreme privation since 2006. As Yahya Sinwar, the assassinated Hamas leader put it, reflecting on his life after a twenty-two-year sentence in Israeli prison: "I came out, and I found the internet. But to be honest, I never came out—I have only changed prisons. And despite it all, the old one was much better than this one.

---

10 Patrick Wolfe, "Settler Colonialism and the Elimination of the Native," *Journal of Genocide Research* 8, no. 4 (2006): 387–409.

I had water, electricity. I had so many books. Gaza is much tougher."[11]
So the crimes against humanity committed in the wake of Operation Al-Aqsa Flood were not entirely new, they were an extreme
escalation of what Gazans had already been suffering through and
which elders remembered from the Nakba that forcibly displaced
them in 1948. The worst damage was inflicted by Israel's massive
military assault, but many of the genocidal measures were the result
of strategic deprivation, including the blockage of humanitarian aid
of food and water supplies. Gazans were no longer being allowed to
stay alive.

In 2012, COGAT was legally forced by Gisha, an Israeli human
rights group, to make public a confidential study, known as the "red
lines document," detailing how officials had been calculating the minimum caloric intake needed to keep Gazans from starving.[12] Israeli
authorities enforced this "minimum diet" (an average of 2,279 calories
per person per day) through their control of food imports by determining how many trucks of food were needed to maintain this subsistence level.[13] As Dov Weisglas, a government adviser, infamously
explained, "We have to make them much thinner, but not enough to
die," the idea being "to put the Palestinians on a diet, but not to make
them die of hunger."[14] In the years to come, these red lines were routinely crossed when the volume of imports fell short, in some cases
deliberately as a collective punishment for acts of Palestinian resistance. In the wake of Operation Al-Aqsa Flood, the red lines were
erased entirely. The international humanitarian efforts to get aid
trucks into the Strip were systematically thwarted by Israeli soldiers
and civilians alike; the ostensible objective of Israeli's military strategy
of Gaza was to force much of the population to die from dehydration
and starvation. By the summer of 2024, Oxfam was regularly issuing

---

11 Ali Younes, "Sinwar: Another War with Israel Not in Hamas's Interest," *Al
Jazeera*, October 5, 2018, https://www.aljazeera.com.
12 See Gisha's presentation of the document at https://www.gisha.org/UserFiles/
File/publications/redlines/red-lines-presentation-eng.pdf.
13 Amira Haas, "2,279 Calories per Person: How Israel Made Sure Gaza Didn't
Starve," *Haaretz*, October 17, 2012, https://www.haaretz.com.
14 Trude Strand, "Tightening the Noose: The Institutionalized Impoverishment
of Gaza, 2005–2010," *Journal of Palestine Studies* 43, no. 2 (2014): 6–23.

famine warnings, and in September, the UN Special Rapporteur on the Right to Food reported to the General Assembly that Israel's denial of food and water was a form of torture, in addition to a crime against humanity.[15]

These methods of deprivation had little to do with Israel's real security needs, and were not adopted because of any shortage of resources that might otherwise be available to its own citizens.[16] They were an expression of what Achille Mbembe has called "necropolitics," whereby a dominant group maintains its existential power over a racially subordinate population, deciding who will live and who will die. In the case of Gaza, these decisions should be considered in the context of a century-long colonial war against Palestinians that has always weaponized access to basic resources like water, food, and fuel.[17]

The genocidal war prompted a worldwide show of solidarity with the Palestinian people, drawing hundreds of millions of new supporters into the liberation movement, and isolating Israel as a pariah nation, along with its most powerful political allies in Washington, London, Paris, Brussels, and Berlin. In the subsequent backlash to this wave of support, campaigns of repression were launched against pro-Palestinian advocates in Western countries on a scale not seen since the postwar era of the anticommunist crackdown.

THE FIGHT OVER NUMBERS

The Middle East is one of the planet's regions where human habitation is a challenge, and where the legacy of Malthusian fear has fue-

---

15 Michael Fakhri, "Starvation and the Right to Food, With an Emphasis on the Palestinian People's Food Sovereignty," A Report of the Special Rapporteur on the Right to Food, UN General Assembly, Seventy-Ninth Session (July 17, 2024).
16 Some of the organizations focused on water crisis are Alliance for Water Justice in Palestine (https://www.waterjusticeinpalestine.org); the Right to Water Campaign by Stop the Wall (https://stopthewall.org/right2water); and Anera, International Development & Relief Foundation (https://www.anera.org/what-we-do/water).
17 Rashid Khalidi, *The Hundred Years' War on Palestine: A History of Settler-Colonialism and Resistance 1917–2017* (New York: Metropolitan Books, 2020).

led a competition over stingy natural resources. It hosts the world's fastest growing populations in locations where the heat is formidable, water is insufficient, and agriculture is a strenuous undertaking. As climate change intensifies, how can these rates of growth be sustained under conditions of resource scarcity?

Over the past century, Israel/Palestine has seen one of the largest build-ups in population, in spite of the forced exile of many Palestinian residents to other countries after 1948 and 1967. Israel's current population is 9.3 million (of whom two million are Palestinian), with a fertility rate hovering around 2 percent. Its Orthodox [*Haredim*] population, with families of six or more on average, is the fastest growing, with Bedouin households in the Negev Desert [al-Naqab] not far behind. The population of Gaza and the West Bank is 5.5 million, with a fertility rate of around 3.44 percent. These numbers are subject to intense scrutiny, not because of natural resource deficits, but because the doctrine of Jewish supremacy has long been tied to numerical dominance. Fear of losing a demographic majority in the lands between the river and the sea is a surefire trigger for nationalist panic, and for overtly racist efforts to stem increases in Palestinian numbers.[18] Israel's control over entry and exit in the Occupied Territories and the registry of Palestinian births and deaths allows its elected officials to closely monitor, and obsess about, all aspects of population growth.[19]

In 2022, Arnon Soffer, Israel's leading demographer-alarmist, announced that Jews were now a minority, accounting for only 47 percent of the population of Israel/Palestine after noncitizens (mostly migrant workers) were factored in.[20] Right-wingers were guaranteed to lose their minds over such estimates, but in this instance, the

---

18 One infamous example was the Koenig Memorandum, penned by a government official in 1976, which sounded the alarm about Palestinian population growth in the Galilee region, and suggested methods of diluting the numbers and the concentration of Arabs. See https://www.palquest.org/en/historict-ext/34011/koenig-memorandum.

19 See Elia Zureik in *Surveillance and Control in Israel/Palestine: Population, Territory and Power*, ed. Elia Zureik et al. (London: Routledge, 2010).

20 TOI Staff, "Jews Now a 47% Minority in Israel and the Territories, Demographer Says," *The Times of Israel*, August 30, 2022, https://www.timesofisrael.com.

hysteria was accompanied by racist laws. Earlier that year, the Israeli Knesset approved the *Citizenship and Entry into Israel Law*, which effectively banned the unification of Palestinian families by prohibiting residency or citizenship status for Palestinians from the West Bank and the Gaza Strip who are married to Palestinian citizens of Israel. The measure was explicitly adopted to ensure a Jewish demographic majority in the Israeli ethnostate, since lawmakers cited, as a rationale, the 2018 *Jewish Nation-State Basic Law*, which constitutionally enshrines Jewish supremacy in Israel.

The 2022 projection by Soffer, who is known by his detractors as "Arnon the Arab-Counter," was the latest in a series of estimates, dating back to his 1987 study that conjured up the threat to Israeli Jews of a Palestinian "demographic time bomb." Allegedly, these reports helped to persuade Ariel Sharon to withdraw Israeli settlements from Gaza in 2005. "Disengaging" from the Strip removed two million Palestinians from the Israel/Palestine demographic record. According to the new calculus, Israel could still maintain a Jewish majority even if it annexed the West Bank and granted citizenship to all its Palestinian residents. But Soffer was not comforted by the future scenario presented by disengagement. In a May 2004 interview with the *Jerusalem Post*, a month before the Knesset approved the measure, he predicted that "when 2.5 million people live in a closed off Gaza, it's going to be a human catastrophe. Those people will be even bigger animals than they are today, with the aid of an insane fundamentalist Islam. The pressure at the border will be awful. It's going to be a terrible war. So, if we want to remain alive, we will have to kill and kill and kill. All day, every day," and "if we don't kill, we will cease to exist."[21] For Soffer, and those who respond favorably to such exterminationist incitements, the pathway to genocide as a guarantee of demographic superiority was laid out quite clearly long before October 2023. "The only thing that concerns me," he added in the interview, "is how to ensure that the boys and men who are going to have to do the killing will be able to return home to their families and be normal human beings."

---

21 See Ali Abunimah's analysis in "The Gaza Massacre is the Price of a 'Jewish State,'" *Electronic Intifada*, July 25, 2014, https://electronicintifada.net.

For their part, Palestinian nationalists have long provoked Zionist demographic anxiety by promoting Arab natality; Yasser Arafat, among others, used to boast that "the womb of the Palestinian woman" is the "strongest weapon against Zionism." Recruiting reproductive power as a paramilitary auxiliary in this way relegated Palestinian women to a traditional role as mothers, whose contribution to the liberation movement would be to bear and raise children, or to be honored as a mother of a martyr [um al-shahid]. It was consistent with the view, also expressed by Arafat, that gender justice would have to wait until the primary battle against the colonizer was over. A more gender-neutral provocation was the 1964 verse of one of Mahmoud Darwish's best known poems, "Identity Card":

> Write down!
> I am an Arab
> My identity card number is 50,000
> I have eight children
> The ninth will come after a summer
> Will you be angry?[22]

One of the origins for this demographic face-off can be traced to the struggle over land that led up to the Nakba of 1948. Under the British Mandate, the 1937 Peel Commission offered a version of partition for a Jewish and Arab state in Palestine. As part of the plan, more than 200,000 Palestinians would be uprooted and transferred to the new Arab state. An even greater population transfer was envisaged under the UN's 1947 Partition Plan for Palestine, which recommended the creation of independent Arab and Jewish States, comprising 42 percent and 56 percent of Mandatory Palestine respectively (the area around Jerusalem and Bethlehem was to be an international zone). Though the settlers in the Jewish Yishuv were only one-third of the population and owned only 8 percent of the land, they were allocated the majority of the territory, including the most fertile agricultural land. The partition proposal was unacceptable to Arab notables. It was welcomed by Zionist leaders, even though David Ben-Gurion pointed out:

---

22 *Mahmoud Darwish: Selected Poems*, trans. Fawwaz Tuqan and Ian Wedde (Cheadle: Carcanet Press, 1973), 24.

In the area allocated to the Jewish State there are not more than 520,000 Jews and about 350,000 non-Jews, mostly Arabs. Together with the Jews of Jerusalem, the total population of the Jewish State at the time of its establishment will be about a million, including almost 40 percent non-Jews. Such a composition does not provide a stable basis for a Jewish State. This fact must be viewed in all its clarity and acuteness. With such a composition, there cannot even be absolute certainty that control will remain in the hands of the Jewish majority. . . . There can be no stable and strong Jewish State so long as it has a Jewish majority of only 60 percent.[23]

In the premeditated program of ethnic cleansing that preceded the British withdrawal in 1948, and that escalated thereafter, Zionist militias tried to make good on the vows of Ben-Gurion and others to take as much land and to transfer as much of the Arab population as possible. After the expulsion of more than 750,000 Palestinians, only 156,000 remained within the Green Line drawn by the 1949 Armistice Agreements between Israel and Egypt, Lebanon, Jordan, and Syria. The demographic majority won by conquest far exceeded Ben-Gurion's 60 percent. Every subsequent Israeli land grab has been conducted according to the principle of acquiring maximum territory with the minimum Palestinian numbers—all with the goal of preserving or improving on that favored population ratio. In that spirit, the most commonly mooted plan for annexing the West Bank involves the acquisition of sparsely populated territory in Area C and the Jordan Valley, further isolating the 165 "islands," scattered in a noncontiguous fashion, where most Palestinians are concentrated.[24]

The historic progression of land theft is clearly illustrated in the series of maps that show the loss of Palestinian territory since 1948. These are best accompanied by hydrological maps of water flows, which help explain how and when the acquisitions made sense. Without water to "redeem" the land or to service settler colonies, the land

23 Nur Masalha, *Expulsion of the Palestinians: The Concept of "Transfer" in Zionist Political Thought, 1882–1948* (Washington, DC: Institute of Palestinian Studies, 1992), 178.

24 Qassam Muaddi, "Israel's Leaked Plan for Annexing the West Bank, Explained," *Mondoweiss*, June 27, 2024, https://mondoweiss.net.

grabs would have been unsustainable. From the outset, that is why the need to secure an adequate supply was fundamental to the project of Zionist settlement in Palestine. In Theodor Herzl's seminal Zionist novel, the *Old New Land*, the future of his imagined Jewish homeland depends on the heroic efforts of hydraulic engineers. Ben-Gurion, who would be the national leader, "talked about water all the time," according to his peers, and Israel's third Prime Minister, Levi Eshkol, declared that water was "the blood flowing through the arteries of the nation."

Securing the life-giving waters of the Jordan River was the key to the original settlement efforts of the Palestine Jewish Colonization Association and others. In the lead up to the 1919 Paris Peace Conference and the subsequent San Remo Conference that adopted the spirit of the Balfour Declaration's promise of a "Jewish national home," Zionist representatives lobbied hard, though unsuccessfully, to extend the boundaries of Mandatory Palestine to include the Jordan's headwaters, along with the Litani and Yarmouk rivers. The drawing of the final territorial boundaries created four riparians, Lebanon, Syria, Palestine and Jordan, destined to clash over the river's water, and also to propel the Zionist dream of capturing and diverting much of it far to the south to "make the desert bloom."

The dream of fertile colonies in the south was already alive in advertisements to recruit would-be Jewish settlers to Palestine. The seductive prose about fruitful land lying in wait was redolent of the sales pitches of arid lands in the American West. Leopold Bloom, no less, the protagonist of James Joyce's *Ulysses* (1920), ponders one of these pitches on a flier issued by "Agendath Netaim" (a garbled version of Agudath Netaim, a real colonization association):

> To purchase vast sandy tracts from Turkish government and plant with eucalyptus trees. Excellent for shade, fuel, and construction. Orange groves and immense melonfields north of Jaffa. You pay eight marks and they plant a dunam of land for you with olives, oranges, almonds, or citrons. . . . Your name entered as owner in the book of the union. Can pay ten down and the balance in yearly installments.

Bloom is not moved; his association of Palestine is with "a barren land . . . a dead sea in a dead land, gray and old." "It bore the oldest, the first race," he muses, but now, in his mind, it is little more than "the grey sunken cunt of the world."[25]

Bloom's dismissive reaction to the ad was no doubt colored by the widely circulated descriptions of Palestine—by officials, Christian travelers, and literary notables like Mark Twain—as a desolate, parched wasteland. In reality, much of Ottoman Palestine was well-cultivated, mostly through the communal land system of *musha'a*, which was productive enough, in some regions, to generate surpluses in the form of cash crops, like olives, cotton, and oranges. The jointly owned musha'a lands were everywhere anchored by rain-fed cisterns, from which water was meticulously distributed in timed allocations to individuals or clans for small-scale irrigation purposes. This system of distribution quotas is still extant in West Bank villages like Ein Battir and Artas. Hundreds of ancient springs, whose use was maintained by customary rules, formed the heart of village settlements, and provided water for outlying areas. According to Muslim law, water is a common entitlement, and "the right of thirst" is a paramount principle not to be denied to any human or animal. (Hence the numerous water fountains, or *sabils*, to be found in towns and cities, though few are still in use.) Large-scale, public waterworks supplied the dry city of Jerusalem through the Roman aqueduct that ran from Solomon's Pools, near Bethlehem, or the rock-cut tunnel that brought water from the Gihon Spring in the Kidron Valley. In sum, the barren, unoccupied Palestine awaiting a resurrection that would "turn the desert green" under technologically adept Jewish direction was nothing but a serviceable Zionist myth.

CONQUEST OF WATER

The roots of water apartheid in these lands lay in the first major concession awarded during the British Mandate. Pinchas Rutenberg, a Russian Jewish hydraulic engineer, won the right in 1921 to utilize much of the running water in Palestine to generate hydroelectric power. His Palestine Electric Company began to supply electricity to Jaffa and Tel Aviv from facilities built on the Yarkon River [*al-Auja*]

---

25 James Joyce, *Ulysses* (London: Penguin, 1992), 60–1.

and in the Jordan River basin just to the south of where the Yarmouk River flows into the Jordan. Rutenberg was not simply an entrepreneur; he was a founder of the Jewish Legion, an auxiliary of the British Army in the war against the Ottomans, and also of the Haganah, a Zionist paramilitary organization that would lead the colonial war against Palestinians. He later served on the Jewish National Council. The award of the water and electric concession to a leading Zionist, who employed only "Hebrew labor," was rightly seen by Palestinians as a preferential move by the British authorities. It was one of the many acts of favor that prompted the Great Arab Revolt from 1936–39, an anticolonial insurgency waged against British policies perceived to be aiding and abetting Zionist settlement.

While the long revolt was ruthlessly put down by British armed forces, the Mandate authorities sought to placate Palestinian concerns by revisiting the explosive issue of Jewish immigration. The principle of "economic absorptive capacity" had long been a factor in determining the Jewish immigration quota, as cited in two White Papers, in 1922 and 1932. These estimates were based, in part, on surveys of the available water supply. Most controversially, the principle of capacity was deployed in the 1939 White Paper's assessment that Palestine's natural resources, including its water reserves, could only support a population of two million people, or 500,000 more than the existing population. As a result, the paper recommended limiting Jewish immigration to 75,000 annually for five years, after which the Arab majority would have a say on any further immigration. This scheme would have effectively preempted any Jewish majority in Palestine. Zionist pushback on these estimates and on the immigration restrictions took many forms, but one was based on the appeal to Jewish ingenuity and willpower to stretch and exploit all water resources to support a much larger population. Ben-Gurion vowed, "No square inch of land shall we neglect; not one source of water shall we fail to tap; not a swamp that we shall not drain; not a sand dune that we shall not fructify; not a barren hill that we shall not cover with trees; nothing shall we leave untouched."[26]

---

26 David Ben-Gurion, *The Peel Report and the Jewish State*, Vol. 10 (London: Palestine Labour Studies Group, 1938), 63.

A key Gentile voice in this pro-immigration campaign was that of Walter Clay Lowdermilk, a soil conservationist whom the US Department of Agriculture sent to survey land and water use in the Middle East. Lowdermilk was impressed by the record of land reclamation on the part of Zionist colonies, and in his report (which became an immensely influential book, *Palestine, Land of Promise*), he argued that the land was capable of sustaining many more people, including four million more Jews.[27] The American efforts at land reclamation in the far West, especially through the damning and diversion of the Colorado River, had shown how water could be stored and moved though inter-basin transfers on a monumental scale. In particular, the recent success of the Tennessee Valley Authority in facilitating irrigation and generating power encouraged him to propose a Jordan Valley Authority. In principle, the scheme would have benefitted Palestinian as well as Jewish farmers, though it implicitly made a case for increased immigration of the latter, whom he believed were much better stewards of the land.

Lowdermilk was also a Christian Zionist who fervently believed that an independent Jewish nation was divinely ordained, in accord with biblical prophecies. His rising reputation was exploited by American and British Zionists seeking to lobby elected officials and elites to push back against Mandatory Palestine's restrictions on Jewish immigration. Opponents in both countries roundly dismissed the campaign and Lowdermilk's findings as "Zionist propaganda," but the heated debate helped to prompt support for the UN's Partition Plan, which would grant an independent Jewish state with its own power over immigration.[28] In the meantime, the World Zionist Organization had commissioned a TVA engineer, James Buchanan Hays, to flesh out Lowdermilk's Jordan diversion scheme, and his plan (published in January 1948) was a detailed blueprint for what would become Israel's National Water Carrier (NWC) project.[29] Indeed, one

---

27 Walter Clay Lowdermilk, *Palestine, Land of Promise* (London: Victor Gollancz, 1945), 122.

28 See Rory Miller, "Bible and Soil: Walter Clay Lowdermilk, the Jordan Valley Project and the Palestine Debate," *Middle Eastern Studies* 39, no. 2 (2003): 55–81.

29 James Buchanan Hays, *T.V.A. on the Jordan: Proposals for Irrigation and Hydroelectric Development in Palestine* (Washington, DC: Public Affairs Press, 1948).

of Israel's first formal acts as a state was to seek to implement Hays' plan. Jordan refused to cooperate with its proposal to irrigate the Jordan Valley, and so the actionable part of the plan was a massive infrastructural effort to divert the water, conveying it over distances of up to 130 kilometers to irrigate the Mediterranean Coast and al-Naqab.

Legitimacy for this massive out-of-basin transfer also came from an American source. In 1953, President Eisenhower sent Eric Johnston, a businessman and former president of the Chamber of Commerce, as a Special Ambassador to broker the division of the Jordan's waters between the four riparian countries.[30] Initially at least, part of Johnston's mandate was to help Nakba refugees settle as farmers on irrigated Jordan Valley land. But Israel lobbied to cut out the refugees and won the right, under the resulting Jordan Valley Unified Water Plan, to pump water out of the river, below sea level, and down the coast to Tel Aviv—a rise of 250 meters—and from there another 100 meters up to Jerusalem.[31] Israel's diversion eventually lifted out much more water than Johnston's plan had stipulated. From 1953 to 1964, the state, with financing from American Jewry, built an extensive network of canals, tunnels, reservoirs, pumps, and pipes to form the north-south backbone of the NWC, with laterals spreading out to the east. Loans from the US government allowed the Israelis to capture and channel large quantities of the groundwater flowing out from the West Bank's aquifers, and from the Coastal Aquifer, into the NWC. The massive public and private investment in NWC infrastructure made the completion of the NWC Israel's primary nation-making achievement.[32] Over time, the grid would be extended much further east, to service the "frontier" settlements in the West Bank, especially in the Jordan Valley.

---

30 Miriam R. Lowi surveys the history of the disputes between riparians in *Water and Power: The Politics of a Scarce Resource in the Jordan River Basin* (Cambridge: Cambridge University Press, 1993).

31 Christopher Ward, Sandra Ruckstuhl, and Isabelle Learmont, *The History of Water in the Land Once Called Palestine: Scarcity, Conflict and Loss in Middle East Water Resources* (London: Bloomsbury, 2018).

32 Seth M. Siegel tells the nationalist story in *Let There Be Water: Israel's Solution for a Water-Starved World* (New York: Thomas Dunne Books, 2015).

In response to these diversions, the Arab League decided to implement its own water weapon—an out-of-basin diversion of two (Hasbani and Banias) of the three headwaters of the Jordan River before they reached the Sea of Galilee, thereby reducing the NWC's draught by 35 percent. Israel responded to this move by bombing the fledgling Syrian facilities in April 1967, hastening on the Six-Day War just two months later. In the aftermath of the week-long hostilities, the Israelis found themselves in control of the entire Jordan River basin and its watershed, draining an area of 1,800 square kilometers, along with the Mountain Aquifer (which consists of three aquifers—the Western, North-Eastern, and Eastern). The West Bank's water was nationalized as state property, in violation of the Hague Regulations of 1907, prohibiting a military occupier from appropriating the natural resources of an occupied population. From that point onwards, all available water sources were carefully audited, and permits were required to drill even shallow wells. Israel was now in total command of Palestinian access to water, including the all-important wells that drew groundwater from the most productive aquifers. Today, as much as 40 percent of its water supply originates as rainfall in the West Bank.

Most importantly, dominion over the West Bank opened the way for Jewish settlement, for which water security and water provision became as important as the NWC had been for population growth within the Green Line. Settlements began as early as 1968, under a Labor government, but the most significant expansion occurred after the new right-wing Likud administration sought to implement the Drobles Plan (or the "Master Plan for the Development of Settlements in Judea and Samaria") proposed by the World Zionist Organization in 1978. Any program of mass settlement would mean that the Occupied Territories had to be integrated with the Israeli water grid. As a result, Palestinian towns and villages were introduced to piped water (some of which had traveled hundreds of kilometers from the Jordan basin) for the first time, but at the cost of ceding their local autonomy. West Bank residents were now dependent on Mekorot, the Israeli water authority, which supplies 58 percent of the West Bank's water, and vulnerable to collective punishment in the form of quota reductions.

Integration of the Occupied Territories into the national water system was promoted to the international community as a benefit of the Occupation, but it was also the clearest indication that they would someday be annexed, either partially or wholly. The capacity to deliver water securely to settler locations far into the West Bank, or to permit residents to drill deep wells nearby, would eventually facilitate the seizure of large areas of land. Many of the pipelines destined for Jewish settlements also supplied Palestinian towns and villages en route, but the lateral pipes servicing the Palestinians were often smaller in diameter, as were the storage reservoirs. Discrimination was therefore built into the infrastructure. The inevitable result was chronic shortages for the villages and towns during the dry season, and especially the hottest summer months when reservoirs run low.[33] Palestinian users also pay much more for their supply than the heavily subsidized settlers.

Israel has seen its population grow twelvefold since 1948, which makes it a demographic outlier. Its birth rate is much higher than the average for an OECD country. This is largely due to a pro-natalist national policy driven by the belief that Israel has a special responsibility to replace the millions of Jews lost to the Holocaust. As a result, any talk of population constraints, let alone controls, in a country that is among the world's most densely populated, has long been considered a heresy. Of course, the high domestic fertility rate was also augmented by mass immigration. Half of the new immigrants arrived since 1990, including one million Jews from the Soviet Union. Securing enough water to absorb these new arrivals became a national priority, shaping Israel's approach to the critical "water-sharing" arrangements that were agreed upon in Oslo.

The Oslo Accords transferred some operational autonomy to Palestinians at the same time it set new rules about the overall Israeli control of the Occupied Territories, including its water resources. Under Article 40 of Oslo II's Interim Agreement in 1995, the responsibility for distributing potable water was vested in the new Palestin-

---

33 Jan Selby, *Water, Power and Politics in the Middle East* (London and New York: I.B. Taurus, 2003), 70.

ian Water Authority (PWA), while a Joint Water Council (JVC) was established to adjudicate permit applications and other claims. The PWA was granted a quota—the use of up to 118 million cubic meters of water per year—which has remained unchanged despite the growth of the West Bank population. Of the water from the Western and North-Eastern aquifers, 80 percent (483 mcm) was allocated to Israel, leaving only the exploitation of brackish water from the Eastern Aquifer to meet the needs of additional households in the West Bank. By 2009, according to a World Bank report, Israelis were extracting 80 percent more than they were allotted at Oslo, and Palestinians were extracting even less than before they received their quota.[34]

For Palestinians, Oslo II was a disastrous agreement, negotiated by unprepared and unqualified PLO representatives. As an upstream supplier, Mekorot ended up with all the power and none of the responsibility for distribution or for bill-collecting, which is a particularly difficult job given how many Palestinian users resist paying Israel for water that they consider, rightfully, to belong to them. Even if the JVC approves an application for a well, the ultimate decision lies with the military authority of the Civil Administration. As a result, very few permits are ever granted. Not surprisingly, Palestinians are well practiced in the art of disconnecting meters or cutting into local lines to divert water, and clandestine wells are not uncommon in rural areas. As with other aspects of the Oslo Accords, Israel fortified its control under Article 40 while delivering itself from the messy, day-to-day management of household water needs. Like the Palestinian Authority (PA) itself, the PWA was established as a weak client, and it has mostly functioned as a conduit to secure donor capital for funding improvements to the notoriously leaky water infrastructure.

In August 2023, I went to the PWA's Al-Bireh headquarters to talk to Subhi Samhan, director of research, about the complex landscape of water distribution bequeathed by Oslo. His account was a long string of grievances. "In the summer," he explained, "we can only ever have intermittent delivery, and it is uneven across the West

---

34 *Assessment of Restrictions on Palestinian Water Sector Development*, Report No. 47657-GZ, World Bank (Washington, DC: April 2009), 9–13.

Bank, because we are not even allowed to swap water from the north to the south where they have much less." So, too, he noted that "even though we have a vast supply of groundwater under our legs, we are only allowed to tap into the Eastern Aquifer, where the water is salty and not cost-effective to treat." In the meantime, "settlers are allowed to drill one thousand feet down into the lower aquifer levels, where the best water lies." He especially lamented that the PWA is severely restricted in its ability to treat wastewater and is taxed when sewage flows across the Green Line. The existing system, he concedes, breeds corruption. The private water tankers, a common sight on streets and rural roads, "are unlicensed, and they charge high prices for poor quality, desalinated water." Local governments also take advantage of the shortages. The West Bank, he explains, "used to have as many as four hundred local service providers, now we are down to eighty, but many of them set their own prices and profit from water sales." Shrugging his shoulders, he points out that "all of this is illegal."

Samhan warms up to the old dream, dating to Rutenberg, of generating hydropower across the eastern slopes of the Jordan Valley: "there is a thousand-meter drop between Ramallah and Jericho, which is a 20-to-1 gradient, and that's a lot of electricity." But, he adds, "we cannot get investors to commit unless we have the water guarantees." As the conversation turns to donors, his real purpose with me seems to emerge: he wants me to find funding for a master's program that will enroll Palestinian students. "They will be able to do your water research for you," he assures me. I leave the PWA building, which literally sits on the divide between eastward and westward flowing water, well equipped with investor information.

My next stop was to visit Emil Abdo, the director of the engineering department at the Jerusalem Water Undertaking. One of the largest service providers, his company covers 600 square kilometers of towns and camps (refugees don't pay for water or electricity), including Ramallah and Al-Bireh. "Our supply," he reports, "is 90 percent from Mekorot and 10 percent from our own wells." "The quota has been fixed for twenty years, and even though we are prepared to pay, we cannot get any more, and sometimes we get less—Mekorot decides what it needs and we get what's left over." Abdo bewails the reduced volume of supply during the hottest summer to date: "In the village where I live, the houses at the top of the hill receive tap water only once a month." But he wants to emphasize that "this is not a state of natural scarcity, it is all political, because we could be sustainable if we had proper access to our springs and aquifers."

Abdo also works as a climate consultant, but warns me that "climate change is not a priority around here." "You cannot mention it to villagers as an excuse for shortages, they have enough to worry about." Besides, he adds, "they have a right to water, and they know there is enough for them, if only we had control." His big concern is with the well at Ein Samia, a major source for Ramallah/Al-Bireh and the villages to the north. Dug by the Jordanians in the 1960s, it is still under Palestinian control, but settlers are threatening to take it over. They have been seizing springs all over the West Bank, but taking a well as large as Ein Samia would be hugely consequential.

CLEANSING THE JORDAN VALLEY

Ein Samia lies in a valley located fifteen kilometers northeast of Ramallah, near the town of Kafr Malik. Our route there was by no means straightforward, and we were advised, on one occasion, not to drive along a road that, until recently, was safe. Eventually, we came across several pipelines which snaked out of the valley and spread out in different directions. When we got down to the valley floor, it was deserted and eerily quiet. I felt like Jake Gittes in the film *Chinatown*, poking his (soon to be lacerated) nose into the Los Angeles water wars. There was no evidence that anyone was protecting the precious wellhead behind its fenced-in compound. As we later learned, there were a couple of unarmed security personnel inside the compound, but they presented little in the way of resistance to would-be intruders; indeed, a water authority official told me that settlers had recently climbed inside and camped there, no doubt scoping out the facility for future occupation.

The purity and strength of the Ein Samia springs have ensured habitation in this beautiful valley for around seven thousand years. The sunbaked land is strewn with the remnants of ancient cisterns and aqueducts; valuable archeological artifacts have been found in caves and tombs. If it falls under Israeli control, it might be turned into a site for historical tourism, especially if evidence of ancient Jewish habitation is found. No doubt, swimming pools fed by the spring will feature prominently in ads for the area's newly built resorts. Until very recently, the valley was home to a village of over fifty Bedouin families, most of whom had been displaced from Ras al-Tin to the east in the 1960s; their Kaabneh clan was originally from Be'er-Sheva (Beersheba) in the Naqab (Negev). The Bedouin had increasingly come under harassment from shepherd-settlers and, earlier in the year, their school and homes were dismantled. As we were leaving, we ran into Yussuf, a community member, who agreed to talk with us the next day.

After showing us some of the ancient grave sites, he sat down to recount how the attacks had escalated. "At first," he explained, "they allowed their sheep to roam onto our land, and then began to steal our own sheep and burn our animals' fodder. Then they sent their kids to cause trouble. The soldiers and police arrested our own youth for resisting, locked them up, and issued heavy fines." He acknowledged that "the combination of arrests and fines proved to be the decisive tactic in the end." We spoke to him just after their school was demolished by soldiers; he informed us, "the PA did nothing to help us," and his community of two hundred people was forced to move further up the valley into the township where their herding livelihoods would be much harder to sustain. With their departure, he pointed out that there was now nothing to stop settlers from taking control of the well. Indeed, Emil Abdo himself conceded that what was unthinkable until recently—the complete loss of Palestinian access to the wells—could occur in the near future. He and other officials were already working on a back-up plan if the worst happened, but dozens of villages north of Ramallah/Al-Bireh would be severely impacted regardless.

Ein Samia was closely watched in other respects. Yussuf and his community were high-profile victims of a relatively new form of settler land-theft on the part of roving shepherds. To establish their herding outposts, the settlers—who are mostly militant religious Jews with dusty biblical profiles—allow their herd to roam far and wide in order to lay claim to land grazed by the sheep. Initiated in 2018 by the Amana organization, and now comprising more than ninety outposts, this sheepherding strategy has proved much more effective at seizing land than traditional settlements. Indeed, settler leaders like to boast that it took more than fifty years for the built settlements to win 100 square kilometers, whereas the sheep "farms" have seized control of double that area in only a few years.[35] The original Gush Emunim (Israeli ultranationalist, ultrareligious "Bloc of the Faithful") settlers learned that if they acted first and seized land without authorization, their outposts would eventually be legalized by the state. The sheepherders don't have to wait that long anymore. Once they are established, the Civil Administration, now under the control of settler-evangelist Bezalel Smotrich, connects them to services, including running water, and they are given long rifles by the Israeli military. Soldiers who are called out to curb their violent attacks on Palestinian herders are usually settlers themselves and they end up aiding and abetting the assaults, as they did in Ein Samia.

The sheepherders' aggression is ongoing across the Jordan Valley, in an area with vast stretches of land and sparse population: exactly the kind of territory earmarked for Jewish settlement according to the infamous Allon Plan outlined after 1967 by soldier-statesman Yigal Allon. The displacement of Ein Samia's Bedouins and the seizure of the wells could be a tipping point for the effective loss of the entire region to settler habitation. Even before October 2023, almost 150 square kilometers of land between Ramallah and Jericho had effectively been ethnically cleansed as part of a likely takeover.[36] The residents still living in this region were mostly Bedouin communities with a history of being displaced from other areas. After the war began, more than two dozen new herding outposts sprang up, and settlers were given free rein to rampage.[37] With all eyes on Gaza, aggressive force was applied to the eviction of the remaining hamlets. Every day, attacks, deaths, and demolitions were reported. By the following summer, the dead numbered more than 500, with many more wounded, and around 10,000 West Bankers had experienced detention in Israeli prisons.

---

35 "Plundered Pastures: Israeli Settler Shepherding Outposts in the West Bank," *Yesh Din*, December 2021, https://www.yesh-din.org/en; Shane Bauer, "The Israeli Settlers Attacking Their Neighbors," *New Yorker*, February 26, 2024, https://www.newyorker.com.

36 Oren Ziv, "'It's like 1948': Israel Cleanses Vast West Bank Region of Nearly All Palestinians," *+972*, August 31, 2023, https://www.972mag.com.

37 Tamara Nassar, "Israel Quietly Disappears West Bank Land," *Electronic Intifada*, July 13, 2024, https://electronicintifada.net.

Further down the valley (the Wadi Uja) from Ein Samia, on the way to Jericho, we visited the equally powerful Ein al-'Auja (or Al-'Auja Spring). Aside from serving as a recreational destination, it used to supply water for an area west of the Jordan traditionally known as the "food basket" for the Palestinian people, with plentiful acreages of bananas, citrus, and vegetables under cultivation. In recent years, wells have been dug to divert water for nearby settlements, and in 2022, Israeli authorities confiscated 22,000 *dunums* of land (5,400 acres) surrounding the spring in a clear statement of intent.[38] With the sharply diminished flow, agriculture was no longer as viable; residents have resorted to becoming day laborers in the settlements, and formerly irrigated land is either sold to settlers or expropriated by the Israeli state because it has not been farmed.

For many Palestinian villages, artesian springs like Ein al-'Auja once formed the life-giving heart of the community. Because of their social centrality, they also generated a good deal of folklore, including stories about water *djinns* who dwelled within. Tawfig Canaan, the physician and virtuoso ethnographer, collected many of these stories in his 1922 essay "Haunted Springs and Water Demons in Palestine," which documents the widespread superstitions about the spirits who reside in cisterns, springs, and wells. More often than not, benevolent spirits coexist and are at odds with the evil ones. According to Canaan, villagers cited the struggle between the two to explain a spring's intermittent flows. On our visits to several village springs, we heard some accounts of such vestigial beliefs. In Artas, just south of Bethlehem, a town well-known for its djinns,[39] a resident told us he had seen one in the form of a sheep five meters high rise out of the trellis at the mouth of the town's famous spring. In the hills above Al-Bireh, we heard a related story; people had been seen at night digging for treasure in the vicinity of the springs. The practice, we were told, derives from the belief that travelers often buried their valuables for safekeeping near springs. Since the treasure is supposed

---

38 Steve Hendrix, "The Famed 'Jericho Banana' is Vanishing. Under Israeli Occupation, There's Not Enough Water," *The Washington Post*, September 25, 2020, https://www.washingtonpost.com.

39 See Celia E. Rothenberg, *Spirits of Palestine: Gender, Society, and Stories of the Jinn* (Lanham, MD: Lexington Books, 2004).

to be guarded by djinns or by snakes, digging for the bounty can be a perilous enterprise.

There were other reasons why a visit to a West Bank spring can be dangerous. Belligerence on the part of armed settlers makes approaching them risky. Drastic state action has closed off others. In July 2024, soldiers were filmed filling a village spring with concrete.[40] Blocking access to springs—in addition to forms of sabotage such as shooting holes in residents' water tanks and cisterns—is one of the means that settlers are using to force residents out of Masafer Yatta, a collection of villages in the South Hebron Hills, to the south of the West Bank's largest city.[41] Israeli authorities have arbitrarily designated these semi-desert lands as a military firing zone, and have been demolishing herders' homes.

Destroying water supplies is a new low. It is a measure that targets the very existence of its victims, and is aimed at finally driving residents out and into the urbanized areas where they can no longer graze sheep and goats. Urban markets in the West Bank depend on Bedouin herders to supply meat, and so the evictions here and in the Jordan Valley have far-reaching consequences.

On our visit to Masafer Yatta, we were told that soldiers were now seizing and confiscating vehicles off the roads, so our mobility was limited. Villagers were using tractors to travel around and relying on water truck deliveries from the town of Yatta that could be blocked at any time. We passed several springs whose names had been Judaicized, along with shiny new wellheads that had been established for Jewish settlements in the area. On our route, we visited a community of families who had been evicted from the town of Susiya in the 1980s when the remains of a synagogue were found and the area was subsequently designated as a historical tourist site. They were in a state of siege. Amir Nassar, a teenage resident who told us he had no future, reported that "the settlers came to destroy our buildings, our playgrounds and take away our solar panels, and now

---

40 "Israel Pours Concrete into Well and Destroys Irrigation System in the Palestinian Village of Al-Hijrah, South of Hebron," *B'Tselem*, August 3, 2023, http://www.btselem.org.

41 Basel Adra, "In Hottest Summer Ever, Masafer Yatta Sears from Water Apartheid," *+972*, August 13, 2023, https://www.972mag.com.

they are sending their goats and sheep onto our land. There are no springs left, and so we depend on the tanker to bring water." Mekorot had connected Susiya and surrounding Jewish settlements to the Israeli national grid. "Their places are always green," shrugged Amir.[42]

After October 2023, the attacks on Masafer Yatta communities escalated from acts of physical destruction and deprivation to violence against persons. More and more Bedouin communities packed up and fled to towns. According to Abdallah Abu Rahmeh, director-general of the Colonization and Wall Resistance Commission, "we warned the villagers, if you do not stand with the Bedouin today, tomorrow the settlers will come after you.... The Bedouin were the line of defense for villages; they have fallen. Now settlers are at the edges of villages. They will be inside villages next."[43]

On the other side of the central mountain ridge, where rainfall drains to the west of Ramallah and Nablus, village communities have a much longer history of residence, but almost every one of them we visited had a story about water insecurity and settler encroachment. Ancient water springs that originally attracted human settlement had been seized and taken over. Many others are "stranded" in Area C, making them inaccessible to Palestinians. Another source of dispossession is the Separation Barrier, or Apartheid Wall. By carving out an ample Seam Zone to the east of the Green Line, the Wall's path was explicitly planned in order to cut Palestinians off from some of the most exploitable groundwater in the Western Aquifer.

In other places, we found that spring water that still flowed to rural communities was so contaminated it was no longer considered potable. In Ein Qenia valley, once a supply of drinking water for Ramallah, raw sewage now flowed down the hill from Al-Reehan, one of the city's new northern suburbs. The treatment facility, we were told by officials, was overwhelmed. "Our sheep are sick, even the birds here are sick, and no one wants to buy our produce," com-

---

42 The multiyear assaults on Masafer Yatta are documented in the 2024 film *No Other Land* (directed by Yuval Abraham, Basel Adra, Hamdan Ballal, Rachel Szor), which won several international prizes and was nominated for an Academy Award for Best Documentary.
43 Taylor Luck, "West Bank Settler Extremists Widen Campaign Against Palestinians," *Christian Science Monitor*, January 26, 2024, https://www.csmonitor.com.

plained the teenage son of one of the affected farmers. But the foul-ing of this beautiful valley also reflects a pattern of class domination within Palestinian society itself, illustrated here by the disregard of the newly affluent hilltop people for the peasantry below. While all Palestinians endure the water shortages imposed on them by the Israeli government, they do not suffer equally. "Neither the wealthy people nor the PA have any reason to care about us," said the youth. As a result, what was once a lush agricultural oasis had become a sacrifice zone.

Even in Artas, which has played a venerable historical role in supplying water to Jerusalem from the Roman era onwards, the water from the springs is contaminated and no longer drinkable. Unlike in nearby Battir, a UNESCO site and national showcase for its ancient irrigation management system, Artas has no public sewage system. Resident NIMBYists cannot agree on the path of the pipelines and so local feuding and the weakness of the town council has prevented any resolution. Townspeople continue to rely on unlined cesspits for waste disposal and so the spring water is polluted. In the meantime, the council abides by centuries-old customary rules of water distribu-tion to the town's clans for agricultural use.

Artas' problems are a symptom of a larger condition that Sophia Stamatopoulou-Robbins has termed the "waste siege" of Palestine.[44] Not only has the West Bank become a dumping ground for Israeli industrial waste (a classic instance of environmental injustice), but strict curbs placed on municipal efforts to treat wastewater and gar-bage properly have rendered much of Palestinian-controlled areas A and B into trash-filled environments. The inability to get permission to build treatment plants ensures that the majority of Palestinian waste spills out in a raw form into the aquifers or flows down the valleys. When it crosses the Green Line, inevitably, the PWA is taxed for Israeli sewage treatment services under the "polluter pays" princi-ple. Challenged to find adequate fresh water, communities across the West Bank are not able to reuse the quantities they can access, and so they are forced to contaminate their own land as a result. The group

44 Sophia Stamatopoulou-Robbins, *Waste Siege: The Life of Infrastructure in Palestine* (Stanford: Stanford University Press, 2020).

psychology of this form of self-degradation is surely an advanced form of colonialism, whereby the devaluation of the occupied population is so internalized that it results in their despoliation of their own environment.

UNNATURAL SCARCITY

Villages may be in the worst straits, but the West Bank's major population centers have also been subject to the Occupation's imposition of water austerity. In the drier south, many Hebronites are used to having tap water only once a month in the summer, and, when Mekorot ordered a seemingly arbitrary reduction of 25 percent in 2023, even the conservatively minded leaders of the city's traditional family clans were readying to protest in the streets. In 2020, a similar mobilization succeeded in restoring the water ration to the city. Hebron, which hosts a quarter of the West Bank's population, sits on two aquifers, but drilling wells, as elsewhere, is prohibited, and any such activity is monitored by Israeli drones.

The most visible indication of Palestinian residence in Hebron or in any West Bank city is the presence of multiple thousand-liter water tanks on the rooftops of buildings. You will not see these in Jewish urban centers. The ubiquitous plastic containers get filled every time running water is delivered, and households organize their lives around its arrival. Ramallah and Al-Bireh, which have an ample annual rainfall of 615 millimeters (24 inches) and yet struggle to supply their residents, delivery comes once or twice a week in the summer—three times if you are really lucky or if you live (as I was told) near important PA officials. Once the roof tanks are full, they are supposed to cover needs for the rest of the week, or month. But many larger households run out and have to buy at inflated prices from the private tankers or from the bottles on supermarket shelves.

Distributions of the bulk supply from Mekorot are supposed to be determined by estimates of need, but in practice, it also depends on how far you live from the pumping source and at what elevation. Frustration at the uneven supply is typically directed at local service providers or the PWA, as was the crafty intention behind the Oslo agreement, but Mekorot is the one with its hand on the tap, and it can turn it on and off at will. In some ways, that state-owned water company is playing a role similar to how the Histadrut, Israel's national federation of trade unions, operated in the early twentieth century. Never simply an advocate for labor (Golda Meir described it as "a great colonizing agency"),[45] the Histadrut used its heft and donor funding to "conquer the labor market" on behalf of Jewish workers. Mekorot is presiding over an even more existential conquest. Control of water, after all, could not be more central to colonial land acquisition and depopulation. That is why, for Israel, its monopoly over the water supply was such a key part of the Oslo Accords, and why state planners have always paid such close attention to the hydrogeology of watersheds and aquifers. That is also why Mekorot will not renegotiate the 15 percent Palestinian share of available water set out in the fateful Oslo agreement even though the PWA is willing to pay for more. Profit takes a back seat to the project of expropriation that lies at the heart of Zionist policy of land maximization.

Nor is it accurate any longer to speak of an "ultimate scarcity." After an acute drought in the late 1990s, Israel decided to develop large-scale seawater desalination plants. Up to 85 percent of the country's potable water is now produced through these plants and the vast majority of its sewage is recycled at a potable standard, though it is primarily used for irrigation. As a result, Israel has such a significant water surplus that it has been selling tens of millions of cubic meters to Jordan every year. In 2023, as part of the Abraham Accords, it entered into a new agreement, brokered by Masdar, the UAE's clean energy company, to provide 200 million cubic meters of desalinated water to the parched kingdom in return for solar energy produced by a large 600 MW solar farm.[46] The agreement, which broke down as a result of Amman's outrage at the genocidal war, was a raw deal for the Jordanians whose acute desperation for water was being traded for continued reliance on Israeli gas exports for their energy needs.

---

45 Tony Greenstein, "Histadrut: Israel's Racist 'Trade Union,'" *The Electronic Intifada*, March 10, 2009, https://electronicintifada.net.
46 "Israel Advances Water-for-Energy Deal with Jordan, UAE," *Al Jazeera*, September 26, 2023, https://www.aljazeera.com. Talks at COP28 to finalize the arrangement, called Prosperity Green, broke down as a result of the Gaza genocide.

In return for the water, Israel would have been able to extract clean energy from its weaker neighbor to fuel its desalination plants.[47]

However exploitative, the stalled agreement highlighted that fact that the Israelis have water to spare and can produce much more with their technologies. In principle, there is nothing to prevent Israel from entering into a similar water-for-solar-energy trading arrangement with the PA. The Jordan Valley, after all, is an ideal location for solar farms. Instead, Palestinians are denied permits to construct solar panels, and they are confiscated or demolished if they manage to do so.[48] As one of our Ramallah interviewees put it, "you really have to be depraved to deny us the use of elemental things like sun and water."

The disparities in West Bank water provision are indisputable and are easily characterized as a form of apartheid. But the use of this label by journalists or NGOs has not done much to slow the grinding machinery of the Occupation or the relentless advance of settlement—let alone pave a way for Palestinian self-determination. Nor does the apartheid epithet capture the omnipresent trauma generated by loss of Palestinian land and life. Even before the genocidal assault on Gaza, there was a clear need to find a more adequate language to describe what is unfolding on the ground; one that does not shrink from calling out the pitiless violence of settler colonialism.

Viewing the Occupation through the lens of water weaponization is a start. The evidence suggests that Israel is actively *dehydrating* strategic portions of the Palestinian population. Water deprivation is a prime military tactic in the "battle for Area C," the portion of land administered by Israel which comprises 62 percent of the West

---

47 Manal Shqair has described the arrangement as an example of "green colonialism." See "Arab-Israeli Eco-Normalization: Greenwashing Settler Colonialism in Palestine and the Jawlan," in *Dismantling Green Colonialism: Energy and Climate Justice in the Arab Region*, ed. Hamza Hamouchene and Katie Sandwell (London: Pluto Press, 2023), 67–87.

48 "COP26: Destruction of Solar Panels in Area C of the West Bank: An Attempt to Undermine Palestinian Development of Sustainable Energy," *Al-Haq*, November 4, 2021, https://www.alhaq.org; "The Sun Belongs to Everyone: Israeli Demolitions and Confiscations of Solar Panel Aid Projects in Area C of the West Bank," *Al-Haq*, 2018, https://www.alhaq.org.

Bank's land but houses only 5 percent of its population. The strategy is to parch these residents, with a view to pushing them into cantons in Area A and B, where they will be within the orbit of the increasingly repressive PA and its crony capitalist allies. Cutting off or contaminating water supply to the enemy is an ancient military tactic, utilized on several occasions on Palestinian lands.[49] During the siege of Jerusalem in 1099, the governor poisoned and blocked the outside wells to deny water to the Crusader army. In 1948, Zionist paramilitary leaders ordered the poisoning of wells in Acre and other Palestinian towns, causing mass infections of dysentery and typhoid.[50] The Pool of Siloam and Gihon Spring in Jerusalem were also contaminated with typhus and diphtheria bacteria as part of their campaign of biological warfare.[51]

Ever since 2006, when the Israeli blockade began, the Gaza Strip has suffered repeated attacks on its water infrastructure as part of military aggression, and repair efforts have been actively obstructed.[52] In May 2021, the last major assault before October 2023, Israeli bombs targeted water pipes, wells, pumping stations, a landfill and a desalination plant, along with sewage systems, sending wastewater into the streets.[53] This massive destruction of civilian infrastructure left 1.2 million people in a state of deprivation, but even this scale of damage

---

49 Mark Zeitoun, "Water Has Long Been Used as a Tactical Weapon in Warfare," *The Conversation*, March 20, 2023, https://theconversation.com.

50 Ofer Aderet, "'Place the Material in the Wells': Docs Point to Israeli Army's 1948 Biological Warfare," *Haaretz*, October 14, 2022, https://www.haaretz.com.

51 Benny Morris and Benjamin Z. Kedar, "'Cast Thy Bread': Israeli Biological Warfare During the 1948 War," *Middle Eastern Studies* 59, no. 5 (2023): 752–76.

52 Erika Weinthal and Jeannie Sowers, "Targeting Infrastructure and Livelihoods in the West Bank and Gaza," *International Affairs* 95, no. 2 (February 2019): 319–340; "Water Supplied in Gaza Unfit for Drinking; Israel Prevents Entry of Materials Needed to Repair System," *B'Tselem*, August 23, 2010, http://www.btselem.org; "By Targeting Water, Sanitation and Hygiene Infrastructure, Israel is Seeking to Render Gaza Unlivable," *Al-Haq*, June 2, 2021, https://www.alhaq.org; "The Right to Water and the War Crime of Pillage," *Al-Haq*, December 10, 2022, https://www.alhaq.org.

53 Marc Santora, Iyad Abuheweila, and Patrick Kingsley, "Hundreds of Thousands in Gaza Face Shortages of Clean Water and Medicine," *The New York Times*, May 18, 2021, https://www.nytimes.com.

was far surpassed by the war crimes committed after Operation Al-Aqsa Flood as part of what the Israeli Defense Forces (IDF) called "Operation Swords of Iron."

HELLFIRE ON GAZA

Even before the war, residents of the Gaza Strip had access, on average, to only about 80 liters of water a day, far below the WHO recommended minimum of 100 liters. Gaza sits on the rapidly depleting Coastal Aquifer and so its residents have some access to groundwater, but the aquifer quality has deteriorated as a result of over-pumping and by nitrate pollution from fertilizers. As a result of the over-drafts, seawater intrusion now extends for as much as three kilometers inland. In addition, Israel dams and diverts the groundwater to service the settlements surrounding the Gaza Strip. As Eyal Weizman observes:

> On the Israeli controlled side are miles of field crops—strawberries, melons, herbs and cabbages—irrigated by state-of-the-art irrigation watering systems and benefiting from the majority of the aquifer waters. . . . To the eyes of the colonizer, looking across from the fences, Palestinian land seem like dead lands. . . . East of the siege lines, the landscape immediately turns dry . . . [Israel has] transformed a once lush and agriculturally active border zone into parched ground, cleared of vegetation, a colonial made desert.[54]

Despite this imposed desertification, and the defilement of the aquifer, more than 80 percent of Gaza's overall supply was still drawn from this groundwater before October 7. Roughly 12 percent, or 18 million cubic meters, came from Israel, sold by Mekorot and delivered through the three pipelines of Al Montar, Bani Suhaila, and Bani Saeed that supply the Governorates of Gaza City, Khan Younis and Deir El Balah, respectively. These were shut off after the Hamas attacks, and only one operated intermittently in the months following, delivering a fraction of the quota.

---

54 Eyal Weizman, foreword to Shourideh Molavi, *Environmental Warfare in Gaza* (London: Pluto Press, 2024), ix–xi.

Ninety-seven percent of the aquifer water was already unsafe for human consumption before October 7.[55] In the bombing raids that followed, water infrastructure was among the specified targets, with wells, boreholes, desalination plants, and wastewater and sanitation facilities selected for destruction. Under the rules of war, combatants are obliged to protect critical civilian infrastructure from damage, but the IDF had crossed that line so many times before that it was barely mentioned in mainstream media, which, in the West, framed its coverage in implicit support of the Israeli offensive. As the weeks went by, and the bombing of hospitals and other medical facilities occurred, it was clear that Israel intended to foment a public health collapse, with inevitable genocidal consequences.[56]

The first symptom of the crisis was the acute shortage of clean water, as the world witnessed images of children lining up to fill jerry cans and bottles. Dehydration preceded the widespread condition of famine. Wastewater flowing through the streets accelerated the transmission of waterborne diseases. Rates of diarrheal illness, along with hepatitis A, skyrocketed; influenza, chest infections, skin ulcers, scabies, and lice spread uncontrollably; and the threats of cholera and polio haunted the devastated towns and urban centers. The struggle for basic personal hygiene became a fight for bare life.

Over the next few months, the IDF's precision bombing took out ever smaller components of the water network, including storage tanks, quality testing labs, and sanitation pumps. Even aid trucks bearing bottles of water were bombed. Israel's military authorities denied or delayed the entry of parts for repairing critical infrastructure for months, and actively obstructed humanitarian efforts to support desalination facilities. A large seawater plant constructed by the UAE in the Sinai provided some relief for a few months until its supply line was ruptured during the Israeli military ground invasion of Rafah. By May 2024, a BBC satellite survey showed that more than half of all Gaza's water facilities had been destroyed or severely dam-

---

55 "Before the Taps Run Dry: Responding to Gaza's Existential Water Crisis," *Anera Reports* 11 (July 2022), https://www.anera.org.
56 Kayem Ahmed, "Israeli Authorities' Cutting of Water Leading to Public Health Crisis in Gaza," Human Rights Watch, November 16, 2023, https://www.hrw.org.

aged.[57] The carnage continued unchecked by any international pressure. In July, IDF forces filmed themselves laying explosives inside a major reservoir serving Rafah; video of the destruction was brazenly posted on social media.[58] By October 2024, all of Gaza's sewage treatment plants had been heavily damaged by bombs and the PWA reported that up to 85 percent of water and sanitation facilities were inoperable.[59]

In addition to water, Israeli authorities had cut off the electricity and fuel supply after Operation Al-Aqsa Flood. The energy grid was subsequently devastated by bombing, and so the only power available to operate water wells and desalination filters came from solar panels. In recent years, Gazans had succeeded in installing a high concentration of rooftop solar systems, supplying 25 percent of daytime energy needs, but most were destroyed in the bombing raids. A July 2024 Oxfam report estimated that as many as five water-related infrastructures had been taken out of service every three days by the bombardment. According to the report, the average Gazan had access to only 4.74 liters of water a day for washing, cooking, drinking, and laundering, far below the fifteen liters cited by the WHO for *basic survival in emergencies*, while a quarter of the population was suffering from waterborne diseases.[60]

Two actions taken by the Israeli authorities stood out for their barbaric intent. The first involved a mass population transfer. Almost two million Gazans were forcibly displaced to "safe zones" where they were promised access to humanitarian aid. Instead, these densely populated locations were subsequently bombarded, including some of the "humanitarian corridors" leading to the zones. In effect, a death

---

57 Kayleen Devlin, Maryam Ahmed, and Daniele Palumbo, "Half of Gaza Water Sites Damaged or Destroyed," *BBC*, May 9, 2024, https://www.bbc.com.

58 Younis Tirawi and Ryan Grim, "The IDF Just Destroyed a Key Rafah Water Facility Rachel Corrie Spent Her Last Month of Life Defending," *Drop Site*, July 28, 2024, https://www.dropsitenews.com.

59 "Situation Report #144 on the Situation in the Gaza Strip and the West Bank," UNRWA, October 20, 2024, https://www.unrwa.org.

60 Lama Abdul Samad, Martin Butcher, and Bushra Khalidi, "Water War Crimes: How Israel has Weaponized Water in its Military Campaign in Gaza," Oxfam, July 18, 2024, https://oxfamilibrary.openrepository.com.

trap had been cynically laid for them, with bait laid in the form of food and water. Initially deployed to divert people to the south, it was then used in North Gaza to accelerate the process of depopulation.

The second action involved the plan to flood the "Gaza metro"— an estimated 1,300 tunnels—with seawater, ostensibly "to neutralize underground terrorist infrastructure" and flush out Hamas operatives. The IDF conducted some trial runs after its ground invasion began in December, and then, at the end of January 2024, officially began to pump water into select tunnels. The news generated outrage about the ecological catastrophe that would inevitably result from saline water flowing into the aquifer. The consequences for any future agriculture in the Strip would be devastating.[61] The UN special rapporteur for the right to water, Pedro Arrojo-Agudo, compared it to the Romans' salting of the fields of Carthage to render their enemy's territory uninhabitable.[62] Egypt had previously used the tactic to eliminate smuggling tunnels on Gaza's southern border. For others, the practice recalled the efforts of the US Army in the 1960s to compromise the tunnels used by Vietcong fighters, along with other programs of biological warfare that contaminated the natural environment of North Vietnam in ways still evident today.

The bombardment of the Strip was "not only a crime against humanity," it was also "a crime against nature."[63] The overall environmental toll taken by the Israeli assault amounts to a sustained act of ecocide. Months of carpet bombing left the land poisoned and depleted, with agricultural production in ruins and unlikely to be restored without massive detoxification efforts. Pollution on such a scale and in such a densely populated place can only result in mass die-offs of people, wildlife, and ecosystems. Climate activists like Greta Thunberg took up the slogan, "No Climate Justice on Occu-

---

61 Ronen Bergman, "Israeli Military Confirms It Has Begun Flooding Hamas Tunnels," *The New York Times,* January 30, 2024, https://www.nytimes.com.
62 Damien Gayle and Nina Lakhani, "Flooding Hamas Tunnels with Seawater Risks 'Ruining Basic Life in Gaza,' Says Expert," *The Guardian,* December 24, 2023, https://www.theguardian.com.
63 David Klein, "Genocide Fuels Climate Crisis. The Fate of Palestine Shapes Our Climate Future," *Truthout,* May 9, 2024, https://truthout.org.

pied Land," but the scope of their protests was dwarfed by the orgy of fossil-fueled violence against Gazans, bearing a colossal "carbon boot print." By April, after two hundred days of continuous bombing, more than 70,000 tons of high explosive (surpassing the bombing of Dresden, Hamburg, and London combined during World War II) had been dropped on ecosystems already struggling to cope with climate stress.[64]

As the carnage from the hellfire mounted, genocidal intent was more and more accepted as an explanation for Israel's actions.[65] Many concluded that the ostensible goal of the Israeli government—to strike a fatal blow to Hamas—was simply a pretext for ridding Gaza of Palestinians, or failing that, for disabling the capacity for reproducing Gazan society for decades to come. The only potential silver lining was that, unlike the "promised land" of Judea and Samaria (aka the West Bank), Gaza is not biblically significant. That it does not figure significantly in the Zionist imagination might have saved it from a large-scale land grab, but the commercial value of its waterfront real estate did not go unnoticed. Jared Kushner was the first to mention this fact publicly in January 2024,[66] all but inviting developers to conjure up blueprints for million-dollar seaside condos. To be built, of course, by Palestinian workers with little to no alternatives.

In May 2024, largely in response to widespread calls for its plan for the "day after," the Netanyahu administration rolled out "Gaza 2035," a more comprehensive blueprint for rebuilding much of the Gaza Strip from scratch, deploying capital from a coalition of Arab countries (Saudi Arabia, UAE, Egypt, Bahrain, Jordan, and Morocco). The AI renderings depicted high-rise urban clusters redolent of Gulf cities like Dubai and Doha, and a transportation and commercial infrastructure designed to position Gaza as an industrial port in a

---

64 Cited in Robert Pape, "Hamas Is Winning," *Foreign Affairs*, June 21, 2024, https://www.foreignaffairs.com.

65 See Forensic Architecture's platform, Cartography of Genocide, displaying the cumulative evidence for this claim, based on detailed mapping of the violence, at https://gaza.forensic-architecture.org/database.

66 David Wintour, "Jared Kushner Says Gaza's 'Waterfront Property Could Be Very Valuable,'" *The Guardian*, March 19, 2024, https://www.theguardian.com.

regional free trade zone (encompassing its Israeli neighbors, Sderot and El-Arish), that would compete with Chinese EV manufacturing.[67] Normalization with Saudi Arabia, anchored by ties with NEOM, its megaproject, lay at the heart of the vision of Gaza 2035, while the professed political goal was to transform the area from "from an Iranian proxy to a moderate hub." Unfettered access to the Gaza Marine gas fields just off the coast was also key to the proposal; the PA is supposed to have sovereignty over its maritime territory, but it has been prevented from extracting these fossil resources which Israel has long coveted.

When Donald Trump took office and announced his development plan for the ethnically cleansed Gaza Strip, the game plan behind "Gaza 2035" generated interest far beyond his far-right Israeli devotees. Indeed, an opinion poll found that 82 percent of Israeli Jews supported the principle of forced relocation and only 3 percent believed the plan to be "unacceptable and immoral."[68] As Yair Wallach pointed out, just as the ethnic cleansers of the Nakba took their cue from a British partition proposal to displace more than 200,000 Palestinians from the Galilee, so too, Trump's backing for the displacement plan would give legitimacy to the otherwise unthinkable idea of evicting two million Gazans.[69]

In the meantime, there was a steady drumbeat from the far right, calling for Jewish resettlement of Gaza and expulsion of its residents.[70] The campaign known as "Returning Home" was launched as early as November 2023, and a high-profile conference called "Security Through Settlement," attended by many cabinet ministers, was convened in January 2024. A festive event, infused by religious messi-

67 Yuval Barnea, "From Crisis to Prosperity: Netanyahu's Vision for Gaza 2035 Revealed Online," *Jerusalem Post*, May 3, 2024, https://www.jpost.com.
68 Jewish People Policy Institute, "JPPI Israeli Society Index, February 2025: The War in Gaza, Social Cohesion, and The Judiciary," February 23, 2025, https://jppi.org.il/en/.
69 Yair Wallach, "Trump's Gaza Takeover Won't Happen. But It Has Already Changed the Face of Israeli Politics," *The Guardian*, February 19, 2025, https://www.theguardian.com.
70 "Israeli Ministers Join Gathering Calling for Resettlement of Gaza," *Al Jazeera*, January 29, 2024, https://www.aljazeera.com.

anic fervor, it featured maps outlining multiple settlements and new neighborhoods with Hebrew names across the Strip.[71] Clusters of would-be settler families formed in the following months, readying to go ahead, with or without government approval. In the meantime, the infrastructure for a more permanent military presence was being built out, especially along broad border zones like the southern Philadelphi Corridor, and a corridor in North Gaza.[72] "Voluntary emigration" was the pathway proposed for Gazans, though few could doubt, after the atrocities IDF had wrought, that removal by force was off the table. As soon as the January ceasefire began in Gaza, Israeli forces launched "Operation Iron Wall" in the West Bank to dismantle and evacuate Jenin refugee camp. The IDF extended its destructive assault to camps in Tulkarem, Tubas, Qalqilya, Nablus, and Jericho, resulting in the largest depopulation effort in the West Bank (with more than 40,000 people displaced) since 1967.[73] In the three weeks after bombing resumed on March 18, the UN reported that at least 100 children were killed or injured every day in Gaza. More than 15,000 had been killed since October 2023.

Another future must be imagined, at a time when it seems most difficult to do so. The concentration camp that Israel and its allies created in Gaza has become not only an open grave for so many Palestinians, but also the burial ground of Western humanism, which has always valued some (lighter-skinned) lives over others. The zeal with which elected governments and institutional elites in the US, UK, Germany, France, Italy, the Netherlands, Australia, and Canada championed the mass killings went far beyond the powerful reach

---

71 Oren Ziv, "Turning Zeitoun into Shivat Zion: Israeli Summit Envisions Gaza Resettlement," *+972 Magazine*, January 30, 2024, https://www.972mag.com; Ishaan Tharoor, "Analysis | Israeli Calls for Gaza's Ethnic Cleansing Are Only Getting Louder," *The Washington Post*, January 5, 2024, https://www.washingtonpost.com.

72 "Israel Planning to Establish Permanent Military Outposts in Gaza, Officer Reveals," *Middle East Monitor*, January 25, 2024, https://www.middleeastmonitor.com; Ruth Michaelson, "Slowly but Surely, Israel Tightens its Grip on Gaza's Lifeline to Egypt," *The Guardian*, July 20, 2024, https://www.theguardian.com.

73 Fatima Abdul Karim and Patrick Kingsley, "Palestinian Displacement in the West Bank Is Highest Since 1967, Experts Say," *The New York Times*, February 17, 2025, https://www.nytimes.com.

of the Zionist lobby. The willingness with which these former colonial powers and dominions supplied the weaponry, and the degree to which the US and the UK increasingly isolated themselves from the rest of the world's nations, was illustrative. On the one hand, it recalled the endurance of the colonial desire, dating back to the resolve of British grandees like Earl of Shaftesbury and Lord Palmerston as early as 1840, to resettle Europe's Jews in Palestine as an outpost of imperial conquest and commerce.[74] On the other, the American and British support was a core assertion that their economic interests were served by what a loyal Israel offered—access to new fossil fields, high-tech innovation, military trade, and expertise in the burgeoning security industry deployed at their own borders. And, in billing his genocidal assault as "the victory of Judeo-Christian civilization over barbarism," what Netanyahu offered in return was a grisly acknowledgment of the white Euro-American ascendancy to which muscular Zionists have always wanted to belong. A large part of the commitment of so many people to stand with Palestine after October 2023 came from the example set by Palestinians themselves— the fortitude of their resistance over the decades, and the record of their resilience in the face of so much suffering. This steadfast resolve to cling to the land no matter what is famously known as "*sumūd*," and nowhere is it stronger than in Gaza. The genocide was an unfair test of sumūd for Gazans; understandably, some of them chose a way out when a pathway to safety overseas presented itself. Many stories about remarkable feats of survival came out of the Strip as residents were forced to improvise under the worst conditions imaginable. Some devised innovative ways of generating electricity for their makeshift tents and care facilities, while others figured out how to process drinkable water from the sea and the contaminated ground, or fashion nutrition from the most meager and unlikely resources.

Education and counseling carried on in the tents after almost every school and university was intentionally blown up. The children of Gaza were disproportionately assassinated or orphaned, yet those who survived, and the hundreds more born daily, were central to all efforts at preserving a lifeline to the future. Mutual aid flourished as

---

74 Andreas Malm, "The Destruction of Palestine Is the Destruction of the Earth," *Verso* (blog), April 8, 2024, https://www.versobooks.com/blogs/news/.

people shared what food, water, and skills they had. Residents of flattened neighborhoods began to rebuild houses with materials salvaged from the debris. Gazans refused to be defined by the intention of their oppressor to reduce their lives to rubble.

But no one should expect exceptional forms of resilience from a people who have been so oppressed for so long. As Malaka Shwaikh put it:

> People who wish to support Palestinians need to stop thinking of us as extraordinary human beings. . . . It is time to stop expecting supernatural 'coping mechanisms' from us. Romanticizing Palestinians, expecting us to show our strength, resilience and patience throughout it all, imposes mythical terms on our experience and our everyday struggles. It obscures our humanity, reduces the depravity of Israeli violence, and ignores other forms of violence, especially the structural violence that we continue to face every day.[75]

Shwaikh's response is an important reminder for solidarity allies who have learned to center Palestinian voices and actions. Calling the oppressors to account is an urgent priority, on a par with the recognition that Palestinians must be able to define and determine their own liberation.

From the shattered landscape of the Strip, a properly decolonial ideology of planetary life will have to emerge. The task of rebuilding Gaza and the West Bank should become a model for human freedom and environmental justice. For people all over the world, a decolonized Palestine has become not simply a solidarity goal but also a paradigm for our own liberation from authoritarianism and exploitation. That is surely why so many of us showed up for Palestine, in the streets, on our campuses, and all over our public and media forums after Operation Al-Aqsa Flood. We organized in the teeth of unprecedented repression from state and institutional leaders in the understanding that the struggle with the Israeli state was as much a fight with our own. That is what is meant by the slogan, "Palestine is everywhere."

---

75 Malaka Shwaikh, "Against 'Resilience," *London Review of Books* (blog), January 23, 2024, https://www.lrb.co.uk/blog/.

RESISTANCE FARMING

The water crisis in the West Bank was worse when I returned in the summer of 2024. Beginning in June, Mekorot reduced supply to the central and southern territories by 40–50 percent, threatening overall food production.[76] More than half of Gaza's farms, which had supplied 30 percent of the Strip's food and exported a large volume of produce to the West Bank, were devastated.[77] More and more farmers were being denied access to their own lands by settler intimidation, destruction of crops, and land grabs. There was no imminent threat of famine in the West Bank, but neither was there any guarantee that the far-right Israeli government would not resort to starvation tactics to drive people off the land at some point in the future, as they were doing in Gaza. Yet, as I would find, Palestinians were trying to reinvent their livelihoods on the very soil that the colonizer craved.

After the Occupation began in 1967, the induction of Palestinian men into Israeli wage labor drew many of them off the land.[78] In some cases, this made it more likely that their household plots would be sold to settlers. Alternatively, as farms ceased production, they could be seized by Israeli authorities under an old, but expedient, Ottoman law that permits the state to claim underutilized land. More and more, the task of farming was left in the hands of women who are now a large percentage (32.3 percent in the West Bank and 46.1 percent in Gaza Strip) of workers on family farms, although the ownership titles are rarely in their names. The agricultural sector is considered the main employer of women, engaging about 20 percent of their overall labor, whether in growing, rehabilitating the land, or marketing and selling the crops.

Beginning in the late 1970s, the Women's Work Committee organized training programs for women in several sectors, including agriculture. Much of its activity was aimed at political mobiliza-

---

76 Awad Al Rajoub, "الاحتلال الإسرائيلي يضغط على الضفة الغربية بتقليص حصة المياه" [Israeli Occupation Pressures the West Bank by Reducing Water Quota], *Al Jazeera*, July 29, 2024, https://www.aljazeera.net.

77 Mohammed Hussein and Mohammed Haddad, "How Israel Destroyed Gaza's Ability to Feed Itself," *Al Jazeera*, July 2, 2024, https://www.aljazeera.com.

78 Leila Farsakh, *Palestinian Labour Migration to Israel: Labour, Land and Occupation* (London and New York: Routledge, 2005).

tion, reflected in the splintering of the committee, during the 1980s, into separate groups affiliated with the main factions—Fateh, PFLP, DFLP, and the PCP. The First Intifada produced a more united, interfactional front that called upon Palestinian women to boycott Israeli goods and participate in the production of homegrown alternatives.[79] The urgent need for locally produced food and clothing prompted a wave of cooperatives involved in home gardening, food processing, handicrafts, animal husbandry and the reclamation of larger acreages of unused or abandoned land. This drive to set up a resistance economy based on a return to the productivity of the land evolved under emergency conditions, but it proved to be one of the slow-burning legacies of the intifada years. Whenever labor was withheld as part of a boycott or when Israel rescinded work permits as a form of collective punishment, the land and the promise of autonomy beckoned, both as a practical alternative and a political cause.

Before October 7, as many as 200,000 Palestinians were employed in Israeli workplaces, laboring mostly in construction and, to a lesser extent, in agriculture. About 150,000 were from the West Bank, where wages are three times lower than they are in Israel, and at least 18,500 were from Gaza, where wages were as much as ten times lower and more than half the population was unemployed.[80] According to the International Labor Organization, an additional 35,000 West Bank Palestinians labored in the settlements.[81] This kind of waged work was a crucial pillar of support for Palestinian households. In 2021, under pressure from trade unions, the Palestinian Council of Ministers raised the minimum wage in Gaza and the West Bank, but it was still too low for most households to survive on unless at least one of their members worked in Israel or in the settlements.

---

79 "When Pickles Become a Weapon: Economy of the First Intifada," Palestine Museum, https://www.palmuseum.org/en; Joost R. Hiltermann, "The Women's Movement During the Uprising," *Journal of Palestine Studies* 20, no. 3 (1991): 48–57.

80 Kav La Oved, "We Must Speak about the Palestinian Workers," February 2024, https://kavlaoved.org.il/en.

81 Report of the Director-General, "The Situation of Workers of the Occupied Arab Territories," International Labour Organization, 110th Session (2022), https://www.ilo.org.

Until October 7, more than 20 percent of Palestine's GDP (or $3 billion, an amount equivalent to two-thirds of the PA's budget) derived from employment revenue earned in Israeli workplaces.

After Operation Al-Aqsa Flood, Israel revoked all Palestinians' work permits. With few exceptions, no one has worked behind the Green Line, and West Bank households and businesses have suffered immeasurably. In spite of the demand from Israeli employers to rehire Palestinians, the Netanyahu administration pushed ahead with an agreement with Indian premier Narendra Modi to import as many as 100,000 guest workers.[82] Cut off now from the waged labor that took them away from their land in the first place, would Palestinians "return to the land," as they had done during the First Intifada or during the pandemic, which saw an upsurge in home gardening? The starvation tactics deployed in Gaza and the rescinding of Palestinians' work permits only reinforced the need for sustainable livelihoods and the provision of basic goods to be independent of Israeli colonizers.

The West Bank teems with examples of sumūd, or resilience, on the land. Even casual visitors cannot help but notice the ancient agricultural infrastructure of stonewall terracing, running all over the rolling hills of the central ridge. Over the centuries, these terraces have preserved soil moisture and quality by limiting surface runoff, while back-up cisterns store water for use during the long dry season. Battir's famously productive terraces are four thousand years old. One well-studied group of traditional farmers are the women who practice the ba'li [rain-fed] system in the Deir Ballut plains of Salfit District.[83] Without the use of chemicals, tillage, or irrigation, their farms have proved sustainable for the best part of a century through crop rotation and the use of heirloom, or landrace, seeds.[84] They do not depend on Mekorot water, and their indigenous techniques stand in marked contrast to the chemical-intensive Israeli industrial farming that supplies most of the produce to be found on Palestinian grocery shelves.

---

82 Andrew Ross, "Migrant Workers in Their Own Land," *New York Review of Books*, April 21, 2024, https://www.nybooks.com.
83 Shqair, "Arab-Israeli Eco-Normalization."
84 Omar Tesdell, Yusra Othman, and Saher Alkhoury, "Rain-Fed Agroecosystem Resilience in the Palestinian West Bank, 1918–2017," *Agroecology and Sustainable Food Systems* 43, no. 1 (2019): 21–39.

Over the last fifteen years, these traditional approaches have been deployed alongside some innovations by a new generation of farmers loosely networked in the name of agroecology, and often under the aegis of the Palestinian Agroecological Forum.[85] They have refined methods for maintaining soil balance by ensuring a high percentage of organic material and beneficial microorganisms, while fertilizing with plant and animal waste. A variety of cropping practices have been tried out, and, of course, water retention through rain harvesting and other means is critical. But for most of the growers, there is also a political principle involved.

Agroecologists subscribe to the concept of "food sovereignty" first introduced by La Via Campesina in the 1990s as the "the right of peoples to healthy and culturally appropriate food produced through ecologically sound and sustainable methods, and their right to define their own food and agriculture systems." The principles were devised as a bottom-up, Global South corrective to the accommodationist idea of "food security," adopted and widely embraced by the corporate agribusiness industry.[86] In a neoliberal food system, adherence to the tenet of food security—associated with the right to import food but not necessarily produce it—has become a formula for reliance on highly processed products from multinational companies. In Palestine, this pattern of dependence is enforced by the Occupation's stranglehold on goods and driven by import profiteers.[87] Because of Israeli restrictions and land seizures, only a fifth of Palestinian land is currently accessible for agriculture.

Among the pioneer farms in the "return to the land" movement were Om Sleiman (in Bil'in), founded by Mohab Alami and Mohammad Abujayyab, and the Humanistic Farm (in Mazare' Al-Nobani),

---

85 Lina Isma'il and Dr. Muna Dajani, "Palestinian Agroecological Forum," Heinrich Boll Foundation, June 2021, https://ps.boell.org/en.

86 La Via Campesina, "Food Sovereignty, a Manifesto for the Future of Our Planet," October 13, 2021, https://viacampesina.org/en.

87 George Kurzom, "Towards Alternative Self-Reliant Agricultural Development," Birzeit University, Development Studies Programme (2001); and the 2021 report by Abdalaziz Al-Salehi, published by the Dalia Association, titled "Palestinian National Food Sovereignty in Light of the Colonial Context," https://www.dalia.ps.

founded in 2010 by agricultural engineer Saad Dagher, an early evangelist for food sovereignty whose many innovations include the creation of a local seed bank for distribution to other community farms. In Beit Sahour, near Bethlehem, Vivien Sansour started a library for cataloging and preserving the seeds of local vegetables and fruit in danger of disappearing entirely. Today, there are as many as thirty agricultural cooperatives in the central and northern West Bank. Advocacy for the movement comes from organizations like Manjala and small NGOs like the Dalia Association, Anera (American Near East Refugee Aid), MAAN Development Center, and the Heinrich Boll Foundation. Since the shared goal is to practice self-reliance, the enterprises are insistent on avoiding the donor trap—when NGOs place stringent conditions on grant recipients. After October 2023, many foreign donors stepped up their threats to withdraw essential aid for organizations that refused to tone down their rhetoric about the genocide or their support for militant resistance.

Alternative agriculture has adherents worldwide, some attracted by the romance of agrarianism or the appeal of healthy lifestyles and others motivated by the urgent need to decarbonize food production and consumption, which accounts for up to 30 percent of global greenhouse gas emissions. In the US, a new generation of young and BIPOC farmers has sprung up nationwide, many of them dedicated to sustainable, chemical-free growing according to agroecology principles.[88] As elsewhere, the cooperatives in Palestine have been founded by college-educated youth. A scarcity of jobs has led them to choose cooperative labor in the fields over nonproductive opportunities in the West Bank's saturated service sector or the humiliations of waged labor in Israel and the settlements.

The Sharaka Youth Forum, convened in 2016 to host workshops involving hundreds of youth, has been an active incubator for the new agroecologists in the West Bank, helping to dispel the condescension usually directed towards those who dirty their hands with the soil. Rami Massad, who coordinates the forum under the aegis of the Popular Art Centre, recalled that, "from our initial sessions, which

---

88 "2022 Young Farmer Agenda," Young Farmers Association, October 14, 2022, https://www.youngfarmers.org.

involved reading books and articles, the concept of cooperatives emerged as the most appropriate fit for the conditions in Palestine." Through resources raised from the community and also the North America-based Palestinian Social Fund, Sharaka was able to help support the formation of the cooperatives by paying for some basic infrastructure: greenhouses, seeds, water tanks, and fencing. Along with "the need to find sustainable employment for youth," Massad is hoping "that the cooperative model will help Palestinians hold on to their land." In other parts of the world, smallholdings are under threat from real estate developers or agribusiness interests. In Palestine, their crops and olive trees are burned by settlers and their land is being seized or confiscated by the colonizer. As Massad points out:

> [T]he protection of land is our real duty. We must do this in concrete ways and not through empty speeches and slogans. . . . How can we talk about a solidarity economy of resistance when we don't have any production? How can we boycott the occupation's goods without producing alternatives? On a social level, the current moment dominated by the PA's policy, and international donors to a lesser extent, promotes a culture of individualism. This is reflected in our current way of life. This is why it's important to rethink a collectivist model, to go back to solidarity and mutual aid. We need to think about collective liberation, and not individual liberation.[89]

During my field visits in 2024, I sat down with Samer Karajah, one of the new generation of young farmers. The charismatic part-owner of a bohemian café in Ramallah, he has two passions. One is *dabke* dancing, which he developed as a youth in a Saffa village troupe called Handala and was subsequently recruited to the renowned El-Funoun company in Ramallah. His other commitment is to Ard Al-Ya's, or Land of Despair, a collective that farms a few parcels near his village, and is central to a network of cooperatives in the region to the north and west of Ramallah. Karajah explained that the name, derived from Nietzsche, speaks to the need felt by many young Palestinians to try

---

89 Nadine Fattaleh and Adam Albarghouthi, "Agroecology, from Palestine to the Diaspora," *The Soil and the Worker* 25, no. 1 (Spring 2022), https://magazine. scienceforthepeople.org.

to transform their alienation into "a productive path towards the future." The first of the Land of Despair parcels was a three-dunum plot which belonged to his family but had been neglected because of its Area C location. A rain-fed crop of peas turned a respectable profit within a year, and Ard Al-Ya's was in business. "I had not really heard about the earlier experiments in agroecology," he confessed, "we just started doing it after we graduated from college because we did not want to work in banks or for corporations, and also because of the value that our culture attributes to the land." They soon added a greenhouse and began to serve their own village community with discounted produce, or through donations to low-income families, bringing the rest of the crop to a newly established Popular Market in Ramallah.

In the summer of 2023, Karajah and some of Ard Al-Ya's members hosted us for an impromptu dinner under the stars on a new 4.5-dunum plot, within a stone's throw of the Israeli Apartheid Wall. They had dug a well and planted a variety of grapes, figs, and almond trees. It was a wild, airy location. "We planned to do dabke dance workshops here," Karajah said, reminding me that dabke originates in labor on the land, through the stomping of feet on the soil to firm it up. Because it was surrounded by settlements, he considered "it was all the more important for us to grow food there." This was resistance farming inside the open jaw of the Occupation. When I returned a year later, it was too dangerous to go to that plot. In the interim, Ard Al-Ya's, which had expanded to sixteen members, men and women, had persuaded an administrator who oversaw *waqf* lands (set aside for religious purposes) to allow them the use of a twelve-dunum parcel nearby. "Do you want to be complicit in the seizure of this land," they said to the official, "or do you want us to use it for our people?"

Influenced by the JADAM organic farming model developed by Youngsang Cho in South Korea, Ard Al-Ya's is now promoting what they call "zero farming," an ultra-low-cost approach that includes homemade fertilizer. On a tour of their facilities, they showed me a cocktail of tomatoes, leaf mold, and water which they had improvised. "Chemical fertilizer is very expensive," Karajah pointed out, "and, besides, the Israelis are restricting it because it can be used for making bombs." After the tour, he was quick to insist that their co-op "is not a romantic effort, we have to show we are economically successful if we want to persuade others to follow suit." This was no easy task; the water shortage and the increasing obstacles to movement around the West Bank posed significant challenges. Most Palestinian farmers were already competing at a disadvantage against subsidized goods from their Israeli counterparts who enjoy discounted water, superior technologies, and economies of scale that permit them to get their produce to market first. Most West Bank farmers—even those with long-established holdings and reputations for quality—were hard pressed to eke out a livelihood from the land alone and often had to seek out additional employment.

We encountered the phenomenon of "weekend farmers" on a field trip to Beit Duqu, a village close to Jerusalem that is one of the claimants to the title of Palestine's best grapes. Villagers used to harvest more than seven hundred tons of rain-fed fruit but were now producing less than half. For market reasons, the leaves of the grapes were now more profitable, though picking them weakens the trees. I noticed that many of the village houses were new and quite upscale. Anwar Sabah, who showed us around, explained that "most of the villagers had other jobs, many in Israel, and only kept up their plots on weekends, farming in a traditional way to respect their inheritance." "We are aware that settlers seriously want this land," he said, pointing to a nearby hilltop where the telltale scars of a settlement and its invasive infrastructure now overlaid the ancient terraces. Before 1948, Sabah's family had grown wheat, cotton, and corn in the town of Lydd, and now he cultivated a dunum of apricots, almonds, and grapes. To be able to maintain a small plot like this was part of the Palestinian Dream, an emotional link to their ancestors.

Others were pursuing the same dream. Samer Mansour, a respected figure in Nablus' Balata refugee camp, drove me down into the Jordan Valley to Al-Nasseriya to see his own recently purchased two-dunum plot of fruit trees, okra, and onions. Before 1948, his family had grown oranges near Jaffa. "This is a way for me to continue the heritage," he admitted, "but it also makes it more difficult for the settlers to think about seizing this land." He would require more water than he had to develop the land further, and so he was content for the time being to "make sure that the land and the fruits stay productive." On a brief visit to Habshe Youssef, an old friend who lives in Bethlehem's Aida camp, I found that he too had recently purchased a small plot to the south of the city. Like Mansour, he had built a small cottage, and planted an array of vegetables—eggplant, tomatoes, and squash. His new neighbors were from Dheiesheh and Azza camps. "They have no interest in producing together, like in a cooperative," he told me, "but we will see." In the meantime, his produce, like Mansour's, was bound for his own household and for his relatives and neighbors in their camps. Neither Sabah, Mansour, nor Youssef had any aspiration to market their crops. Their choices seem to be driven by a mix of agrarian romance and legacy spirit.

Mansour introduced me to his Al-Nasseriya neighbor, Malik, one of five brothers who collectively farmed an ample spread of around 600 dunums (150 acres). As we walked through their crops, he recounted his good fortune: "We have a good water supply, and some of the best produce in this part of the valley," though he conceded that "the new restrictions on fertilizer are a concern." But as he spoke, his eyes strayed to the high ground above us, where we could see the outlines of a makeshift settlement. "The 'shepherds' up there are always coming down to harass us," he said, meaning the roving settlers who started arriving in the area in 2023. "They send drones," he reported, "and they release their wild boars to ravage our fields."

He also told us that on the other side of that hill, there were more than 1,500 dunums of greenhouses and farmed plots, newly erected and tended by Palestinians since October 2023. Even though these new farmers might be in competition with him, he seemed to welcome their cultivation of the land. We heard many other stories along the way about people who had been unable to find paid work and had been trying to stay busy and active by working on their land, or that of their neighbors. Even if their survival diet contained only a small portion of the home-grown produce, it was still a strike against the commercial empire of Israeli-supplied goods.

In many locations, this movement back to the land involved confrontation with hilltop youth or with the expansive actions of established settlements. In the summer of 2024, the Israeli government, under the influence of settler-politician Bezalel Smotrich, moved to annul the Palestinian right to construct and develop in Area B, laying a legal pathway for further house demolitions and appropriation of land near settlements. To grow and distribute the produce consistently as a market alternative would risk further retaliation if it attracted sufficient notice. Every square inch of the Jordan Valley, in particular, is up for grabs. The benign climate and fertile terrain of this natural greenhouse make for a bountiful agricultural prize. It is being ethnically cleansed at a rapid rate, and many of those displaced are recruited as day laborers on the newly established settlement farms.

Settler aggression aside, Palestinian growers who want to hold on to their market share talked about changing the logistics of the producer-consumer relationship. Sabah said he envisaged the formation of grocery collectives who could come directly to the Beit Duqu farms to truck out the fruit. Many of the agroecology farms were pioneering new kinds of community involvement though the local circulation of their produce and the reintegration of agriculture into village life. Whether these initiatives can be scaled up remains to be seen. One thing was sure: there would be no help from the central authorities. The PA leadership pays lip service to the nationalistic value of the land, but directs only 1 percent of its budget to agriculture, while as much 35 percent (as of 2015) is allocated to the security services it performs on behalf of the occupier.[90] Land is central to the neoliberal model fully embraced by the PA since 2007, but it mostly figures in the form of debt-leveraged real estate, not as a productive source of food.

Ard Al-Ya's has fraternal relations with two other cooperatives in the region: Om Sleiman and Ard al-Fallah. The trio call themselves the "West Line," and they regularly trade ideas, techniques, labor, and, on occasion, chickens. Ard al-Fallah's Mohammed Khweireh, another El-Funoun dancer, runs a small farm in his village of Kufr Nimeh, and he is a member of both Ard Al-Ya's and the Ard al-Fellahin cooperative in the village. After he graduated from Birzeit University he "decided to go back to the manure," even though the "goal of a college education," as he put it, "is not to get you dirty." His father is a former *fedayeen* militant in Jordan. His mother has farming knowledge, especially with goats, and so, after a tornado wiped out the first

90 Tariq Dana, "Corruption in Palestine: A Self-Enforcing System," *Al-Shabaka*, August 18, 2015, https://al-shabaka.org.

of Khweireh's vegetable crops, they bought a herd of chamois goats
together. The high price of water in the village was also decisive. "I
prefer the goats," he admitted," they are more sociable, I can talk to
them." Along with 150 chickens, the labneh, milk, and cheese from
the twenty-eight goats are the new base of household revenue, and
he plans to triple the herd's size soon. The produce is distributed to
the village, through the Popular Market or to a few individuals who
have paid upfront for several months of supply.

Like Karajeh, Khweireh is a thinker as well as a doer, given
to reflective analysis of what it means to function as a collective
and to make a community through producing and providing food.
Over lunch one day in Saffa, with several members of both cooperatives, we had an expansive discussion about the Palestinian *fellahin*
tradition, and how new farmers like themselves, college-educated
and politically committed from a middle-class perspective, stand in
relation to that tradition. As small holders whose land was under
threat from Zionist settlers in the early twentieth century, the fellahin constituted the most militant element behind the great revolt of
1936–39. They were fighting to preserve or restore an existing order,
firmly rooted in their landed rights, and were betrayed by the "feudal-
clerical" leadership in the cities.[91]

Today's return to the land is also occurring at a time of escalating
land grabs. By contrast, however, the cooperatives have sprung up
not to preserve the status quo but to forge an alternative to the neoliberal capitalist order introduced by the PA. Our discussion extended
to their aspiration to build a union of cooperatives strong enough to
win and secure rights for the new agriculturalists. In principle, the
economic heft provided by a union or a confederation would help
to sustain even the smallest initiatives. It would also expose them to
Israeli scrutiny and aggression, as with any other independent, or
self-reliant enterprise on the part of Palestinians.

Om Sleiman, the third participant in the West Line cluster,
was founded in 2016 as a CSA farm, the first in the West Bank. It
is located in Bi'lin, a battle-scarred veteran of many years of resist-

---

91 See Ghassan Kanafani's unfinished analysis, *The Revolution of 1936–1939 in
Palestine: Background, Details, and Analysis*, trans. Hazem Jamjoum (New York:
1804 Books, 2023).

ance to the building of the Apartheid Wall on village lands. The farm, whose acreage is owned by a writer in the village and offered gratis, currently has five female members, provides room and board for several volunteers, and produces as many as fifteen crops. Its CSA subscription model serves more than twenty-five households in Ramallah and Jerusalem, and so it does not distribute its produce locally or to the Popular Market unless there is a surplus. Om Sleiman's location, next to the wall and close to a settlement built on stolen Bi'lin land, makes for an especially precarious existence since October 2023. Yara Dowani, the farm's "manager," told me that she and others attract gunfire from soldiers if they venture to the outer perimeter to repair fencing gaps opened up by the wild boars.

College-educated and fresh from an experience with young agriculturists in Spain, Dowani, a Jerusalemite, signed up for Om Sleiman because of a felt attachment to the place and to the spirit of the enterprise. "I did not want to be a farmer," she told me, "I wanted to inhabit a lifestyle." That said, Dowani informed me the farm "answered all of my existential questions about Palestine—economic, political, and ecological." As a result, she said, "I can still be a *fellaha* today, the relationship is unavoidable, not just because of my emotional bond with the land, but because my experience with the seeds, the crops, and the daily tasks is much the same as in the past." Om Sleiman's workers make decisions together. They pay themselves about a thousand dollars a month but typically hold down other jobs. Dowani is aware of criticism from other co-ops that her farm's prices are too high for the villagers. "It's what we need to do to survive as a CSA," she shrugged. "Although we can produce all year around, our prices are still not enough, so we need to look for donations, or we do tours and workshops for extra money."

Dowani did not brag about Om Sleiman, but she knows the farm has set an example, without any help from the PA or from big donors, "We have proved that you can reclaim land and make it productive in Area C, next to the Wall." Palestinians, she noted, "have to fix problems for ourselves—we have no one to complain to—and so we cannot accept being told that we cannot build in Area C." Although her farm was not particularly active in the burgeoning network of co-ops, she pointed out that the increasing difficulty of travel meant that the only effective mode of collective organizing was within the regional clusters of Nablus, Tulkarem, and Ramallah.

On a trip I and my colleagues were advised not to take because of soldier and settler violence on the roads to Nablus, we met with Hakima Jaber, who coordinates the Nablus cluster of about fifteen cooperatives, with several more under formation. Operating out of a center for combating violence against women, she trains women and youth in the basics of running an agricultural enterprise. "Men are sometimes recruited for heavy manual tasks in the fields," she laughed, "but otherwise, we are all women, because women are more cooperative, we have less arguments, and don't need to feel like commanders." Her own co-op is in Asira al-Qibliya village, ten miles southwest of Nablus, in Area B, and it is constantly under assault from some of the most aggressive settlers in the West Bank. "They are criminals from the Israel justice system," she said, "who are sentenced to do 'social service' over here." Among their actions was a vicious assault in March 2023 on residents of the nearby town of Huwara that even the Israeli military authorities described as a "pogrom." Jaber's farm has ten dunums of vegetables, but it revolves around its beehives. In the alternative economy she and others were building, "we practice mutual aid among the co-ops, with in-kind bartering of goods and services."

Like Dowani, Jaber is proud to call herself a fellaha; "I have kept my rural accent and do traditional Palestinian cooking with my *taboun*. I try to lead a simple life." Even so, she was clearly a good organizer and was excited about the new wave of returnees to the land. "We can provide jobs this way," she said defiantly, "and since cooperation on the land is our social heritage, maybe we can regain what we lost after Oslo." She had in mind the communal labor that Palestinians perform around the yearly olive harvest and that sometimes extends to the building of houses.

Even with new cooperatives forming every year, they might not last. Farming is hard work, farming against the grain of the retail market is even tougher, and when your farm is in Area C (or now large parts of Area B), within raiding distance of settlers, your chances of surviving are slim. Yet the co-ops are doing everything right. They are a conscious response to the conditions of the Occupation that include land theft, settler violence, market monopolies, and the corrupt authority of the PA. They are built on the strong memory of agricultural heritage and the promise of newly sustainable practices. Their members adhere to the credo that there can be no liberation without sovereignty over daily bread. The reverberations of this sentiment can be heard on the streets of cities around the world, where protesters chant, "*Eish, hurriyya, Falastin Arabiyya*" ["Bread, freedom, and an Arab Palestine"]. But they are also under a great deal of pressure to succeed. As Massad put it, "if we fail, no one will talk about cooperatives for the next thirty years." This was a high bar to set, even for Palestinians who are used to seemingly insurmountable obstacles, and who, despite their many reasons for despair, remain condemned to hope.

Although the issue of climate change is not a high priority for them, these new farmers are nothing if not practitioners of climate justice. One of their aims, after all, is to liberate themselves from the colonizer's fossil-fuel economy, exemplified by chemically intensive monocropping. Returning their land to productivity can help trap carbon in the soil, and their embrace of crop polyculture is a great boost to biodiversity. Employing more people in sustainable agriculture is also a "just transition" away from dependence on "working for the enemy" in the high-carbon construction industry of Israel and the settlements. Reclaiming their water is a way of ensuring safe and clean supplies to vulnerable communities, and ultimately, may lend itself to the rehabilitation of the formerly dynamic Jordan River, reduced over the decades to a sewage-filled drainage ditch by the damming and diversion of its water for Zionist settlement.

Sunset in Dubai, December 2023.

2

UAE

THE ANTIPODAL TRICKSTER IN DUBAI

Each Conference of the Parties (COP) meeting is a version of Last Chance Saloon. At the annual conference of the United Nations Framework Convention on Climate Change (UNFCCC)—the UN body responsible for combating climate change—even the most apocalyptically inclined have to tell themselves and everyone in earshot that "it's not too late" to turn things around. Despite almost three decades of failing to kick the fossil fuel addiction, there's still time, albeit an ever narrower window, to quit. Too much is riding on the outcome for it all to be just "blah blah blah," in the deathless words of Greta Thunberg.

Because of this quandary, Dubai's hosting of COP28 carried the lowest expectations and potentially the highest stakes. Sultan Ahmed Al Jaber, the UAE's top oil boss, was named president of the climate summit—a sharp slap in the face to those who thought, with good reason, that the world's $4 trillion fossil fuel industry should be taking a back seat. As Dubai readied itself for the proceedings, Al Jaber's

Abu Dhabi National Oil Company (ADNOC), one of the world's largest, brazenly announced plans to grow its production capacity by 7 percent and its capital expenditure by 18 percent over the next four years.[1] A shocking number of lobbyists from the oil and gas industries were issued badges—almost 2,500 by some accounts—each with a seat at a table ostensibly prepared for their last feast. They had come to peddle false or ungrounded climate solutions like "low-carbon" biogas and blue hydrogen, direct air capture and storage, geoengineering, carbon markets, "abated" emissions, or anything else that would allow them to go on pumping.

Dubai's gathering came at the end of a year that smashed through all the planet's temperature records by a huge margin, inaugurating what the UN's own Secretary General dubbed the coming era of "global boiling."[2] Chroniclers of climate change lost count of the number of regional records broken as a result of prolonged heat waves, droughts, floods, and wildfires. Climate scientist Zeke Hausfather memorably described September's soaring global temperatures as "gobsmackingly bananas."[3] Carbon dioxide levels measured at the Mauna Loa Atmospheric Baseline Observatory peaked at 424 parts per million in May, a concentration last seen millions of years before. Scientific experts issued reports about the critical climate thresholds already crossed. In its sixth comprehensive assessment, the traditionally conservative Intergovernmental Panel on Climate Change (IPCC) issued a "final warning" about the need for drastic emissions reductions to limit global temperature rises to 1.5°C above preindustrial levels.[4] If unheeded, the world would have exhausted the remaining carbon budget for that threshold by 2030, the date of the next IPCC report.[5]

---

1   Jamie Ingram, "ADNOC Steps Up Investment as 5mn b/d Plan Brought Forward to 2027," *Middle East Economic Survey*, December 2, 2022, https://www.mees.com.

2   "UN Chief Says Earth in 'Era of Global Boiling,' Calls for Radical Action," *Al Jazeera*, July 27, 2023, https://www.aljazeera.com.

3   Matt Simon, "September's Record-Shattering Heat Was 'Absolutely Gobsmackingly Bananas,'" *Wired*, October 4, 2023, https://www.wired.com.

4   IPCC, *AR6 Synthesis Report*, March 20, 2023, https://www.ipcc.ch.

5   2024 delivered even higher temperatures, prompting many to declare that the 1.5°C cap in the Paris Agreement had already been breached, even if unofficially.

But there were other numbers worth noting. In 2023, the major oil companies posted record profits, guaranteeing that their high rates of return would spur even more investment.[6] In several countries, including the US, Canada, Brazil and Guyana, oil and gas production hit a record high. The US produced more crude oil—13.5 million barrels a day—than any country in history (a fifth of the world's total), continuing an upward trajectory that began with the 2015 lifting of a ban on crude oil exports, while its gas production and exports skyrocketed as a result of its leading role in hydraulic fracking. Worldwide oil and gas upstream capital expenditures saw the largest year-on-year gain ever, and the International Energy Agency forecast that aggregate global demand for fossil fuel would also rise to record levels in 2024.[7]

Despite the stench of petroleum hanging over the proceedings, the COP28 delegates had aspirations to deliver a strong final agreement with language decreeing a phase out of fossil fuels. Dubai's meeting would also be the first to take inventory (through a process called the Global Stock Take) of how much progress each country had made since the 2015 Paris Agreement in reaching its emissions reduction targets. If they fell short, the final agreement was supposed to hold nations to account by setting new goals and attainment deadlines. Inevitably, all parties to the Paris Agreement had failed to reach their domestic targets. Among the worst offenders were major oil producers like Russia, Mexico, Saudi Arabia, and the host nation itself.[8]

Typically, there is a bad guy at each COP, singled out as a scapegoat for the failure to make significant progress. In past years, the United States, China, Russia, and India have all played this role. This time, the smart money was on Saudi Arabia, the most influential member of the OPEC cartel. All eyes were on the meetings being

---

6 Brett Christophers argues that, despite the plunging price of renewable energy, alternatives to fossil fuel have not produced anywhere near the harvest of profits generated by oil and gas industry. *The Price is Wrong: Why Capitalism Won't Save the Planet* (New York: Verso, 2023).

7 "IEA Revises up 2023, 2024 Global Oil Demand Forecasts," *Oil and Gas Journal*, November 14, 2023, https://www.ogj.com.

8 Climate Action Tracker, https://climateactiontracker.org/countries/.

taken by Saudi Energy Minister Prince Abdulaziz bin Salman, who had pushed everyone's alarm bells by publicly opposing any language about a "phase down," let alone a "phase out." Al Jaber, by contrast, was under pressure to deliver, using the persuasive powers of the COP presidency, but he himself was on record proclaiming that a phase out would "take the world back into caves."[9] In the meantime, a leaked report suggested that he had been using the run-up to the conference to pursue new oil deals.[10] It looked as if the fix was in.

Because a climate summit in Dubai was too surreal to miss, I decided to attend my first COP. In 2015, however, I had been barred entry to the UAE,[11] along with three other comrades from the Gulf Labor Artist Coalition,[12] for conducting research on the conditions of migrant workers in that country.[13] By all accounts, it is not easy to get yourself removed from the list of personae non gratae. Still, COP28 seemed like my best shot at returning to the Gulf state. Human rights advocates from Amnesty International and Human Rights Watch (HRW), who had all been banned in 2013, had their visa applications approved for the conference. So, I opted to apply through NYU's accreditation with UNFCCC. I was issued an observer pass, but, unlike other members of the NYU delegation, my visa application was rejected. James Lynch, from the labor rights organization FairSquare, was in the same boat.[14] A *Financial Times* reporter, who planned to file a story about our denials, sent an inquiry to the Dubai visa office. Within an hour, my visa was approved. I flew on short notice to join the seventy thousand professional talkers, dealmakers, activists, and hawkers of green gizmos who filled the vast exhibition grounds built for Dubai's Expo 2020.

9 Laura Paddison, "Climate Summit Leader Defends Controversial Comments," *CNN*, December 2, 2023, https://www.cnn.com.

10 Justin Rowlatt, "UAE Planned to Use COP28 Climate Talks to Make Oil Deals," *BBC*, November 26, 2023, https://www.bbc.com.

11 Stephanie Saul, "N.Y.U. Professor Is Barred by United Arab Emirates," *The New York Times*, March 16, 2015, https://www.nytimes.com.

12 Gulf Labor Artist Coalition, https://gulflabour.org.

13 Andrew Ross (ed.), *The Gulf: High Culture/Hard Labor* (New York: OR Books, 2015).

14 Vivian Nereim, "Protesters Test the Limits of Authoritarian Dubai's Climate Summit," *The New York Times*, December 10, 2023, https://www.nytimes.com.

December in Dubai was unusually hot and sultry, due, no doubt, to global warming. Its property market was also overheated, due to the influx of Russian and crypto billionaires, with prices outstripping London and Singapore. The UAE itself was still riding a boom in "black gold" that began in 2009, refueled now by elevated prices due to the war on Ukraine. And regional politics were on a knife edge, due to the recently launched genocidal war on Gaza. What were the chances that an international summit on the planet's future could alter any of those circumstances for the better?

The conference grounds were split into the Blue Zone, where the delegate sessions and horse-trading would be conducted amid the national and institutional pavilions, and the Green Zone, designated as an exhibition space for businesses and civil society. As a gesture to the COP's tradition of counter-summit protest, some rallies would be permitted in the Blue Zone, which was under international juris-diction. The requirements for approval were draconian, however, and participants, especially Indigenous groups, reported high levels of surveillance, intimidation, and harassment.[15] After many days of haggling, Amnesty International and HRW were allowed to stage a protest about the plight of political prisoners, most notably UAE's premier human rights advocates Ahmed Mansoor, Mohammed al-Roken, and Nasser bin Ghaith, but only if captions from the images detailing the detainees' names and sentences were removed.[16] The country's first pro-Palestine rally since 2009 also took place. By contrast, the much larger Green Zone, whose pavilions resembled a vast trade fair, was entirely UAE territory, and so protests were not only outlawed but also unthinkable.

The disparate vibe of the two zones spoke to the host's inter-pretation of the UNFCCC process. The politicking and lobbying going on in the Blue Zone might have seemed at odds with the frank

---

15 Graeme Green, "Environmental Campaigners Filmed, Threatened and Har-assed at Cop28," *The Guardian*, December 20, 2023, https://www.theguardian.com.

16 Many of these political prisoners were scheduled to be released in 2023, but the UAE laid new terrorism charges against them during Cop28. The Emirates Detainees Advocacy Center condemned the actions, at https://en.edacrights.com/post/672.

commercial promotion and deal-making conducted in the adjacent Green Zone. Yet, as the meetings trundled on to their conclusion, the separation of these two spheres by tightly staffed security barriers—partitioning the sacred from the profane—made more sense to me. The state power of climate diplomacy and the market power of clean-tech were proceeding in tandem, on parallel tracks, but the driving force of fossil power, inconspicuous but unswerving, cut across both at a tangent.

A POST-PETROLEUM STATE?

In many ways, Dubai was an appropriate COP host, because it show-cased all the vices that need to be vanquished if we are going to have anything approaching an egalitarian green society. Fiercely protective of its relative autonomy from Abu Dhabi, its wealthier and more conservative neighbor, Dubai was the last major Gulf principality to strike oil, and the first to exhaust its reserves. Hydrocarbon deposits were found in Bahrain as early as 1932, in Kuwait and Saudi Arabia in 1938, Qatar a year later, and Abu Dhabi in 1958. But Dubai drilled dry or unproductive wells until it got lucky in 1966. But with only 4 percent of the UAE's proven oil reserves, or four billion barrels, its fossil fuel assets were all but gone by the early 2000s, forcing the emirate to diversify by soliciting foreign investments in its burgeoning tourism, finance, construction, and real estate sectors. In the meantime, its ruling Al-Maktum dynasty used the oil proceeds to build out a physical and commercial infrastructure of deep water ports, export-processing free zones, financial trade centers, and light manufacturing clusters that prepared the way for a post-petroleum economy.[17] Its mercurial growth over the next two decades as the preeminent, low-cost trading hub of the Persian Gulf was built on its pre-oil colonial history as a smuggler's paradise, sustained by the rents extracted from pirate-merchants and from its favorable location as a logistics waystation for the British Empire's India trade. Its population doubled between 2001 and 2008, by which time it was estimated that more construc-

---

17 Christopher Davidson, *Dubai: The Vulnerability of Success* (New York: Columbia University Press, 2011).

tion was underway in Dubai than in fast-forward Shanghai.[18] With 3.8 million residents now, its population is expected to surge to 5.8 million by 2040.[19]

Yet Dubai's contemporary trendsetting profile as a self-reliant global city masks its dependence on the UAE's oil resources and military umbrella, reinforced by the heavy US marine presence at Abu Dhabi's Al Dhafra Air Base and a missile defense system provided by Israel. The emirate's zealous commitment to a free market society, leveraged by investor debt and lubricated by hot money flowing from crisis-torn countries, also makes it vulnerable to capitalist crises of overspeculation. With no income tax, property tax or corporate tax, the rentier economy is kept afloat by financial liquidity. When this debt bubble was popped by the 2008 crash, prompting a tsunami of defaults, its reliance on the federation's oil wealth was most fully exposed. The UAE capital, Abu Dhabi, bailed out its entrepreneurial rival with $20 billion in loans and bond purchases, and lent further support during the COVID pandemic.

In the interim, Abu Dhabi, whose heavy manufacturing sector is parasitical on its petroleum assets, bowed to Dubai's diversification model by jump-starting its own tourism and real estate initiatives on the outlying islands of Yas and Saadiyat. The former, with its theme parks, and its racing car, water sports and concert venues, is a recreation and entertainment hub, while the latter is being developed as a chunk of luxury real estate, whose Mediterranean Revival villas are prestigious Gulf addresses with eye-popping prices. Promoted as "a premier island destination" for a projected 125,000 well-heeled residents, Saadiyat (or "Land of Happiness") includes a cultural district boasting brand-name branches of the Louvre and the Guggenheim museums and New York University that are designed by starchitects such as Jean Nouvel, Frank Gehry, Zaha Hadid, Rafael Viñoly, and Tadao Ando. The museums are merely part of the real estate draw— just another amenity for residents who have everything else that

---

18 Jim Krane, *City of Gold: Dubai and the Dream of Capitalism* (London: St. Martin's Press, 2009), 297.

19 Abeer Abu Omar and Zainab Fattah, "Dubai's Allure to Expats Is Weighing on City's Infrastructure," *Bloomberg*, October 13, 2024, https://www.bloomberg.com.

money can buy. The website of Saadiyat's developer makes this crystal clear, noting that the villas "have been built around natural beauty, cultural experiences, architectural splendors, and vast business potential." In this context, access to culture is a value-adding item on a list of selling points, a lifestyle enhancement, and a sales pitch for the swanky homes.[20]

Decadent Dubai does not have the deep pockets to buy the institutional brands, the starchitects, and the art collections to fill such edifices to high culture. Its investor-financed commercial development is geared toward luxury lifestyles, but it has not been dependent on cultural capital. With the bustling feel of an entrepôt that welcomes well-heeled retirees, vacationers, and investors from all corners of the planet, Brand Dubai has become a dynamic model for other MENA states. It is also an environmental nightmare. Sustaining the Gulf's most bling-worthy version of the good life—underpinned by a shadow economy of smuggling, gunrunning, money-laundering, and human trafficking—comes at a high cost to the planet. Dubai's fast-growth fever bequeathed a sprawling, and heavily congested, traffic infrastructure and a forest of soaring glass towers that require voracious energy inputs to maintain. Its car fetishism and love affair with air conditioning is a match for the most unsustainable American corner of Sunbelt suburbia. As a result, air quality along the overdeveloped coast is increasingly compromised and its relationship with the waters of the Gulf is a toxic one.

Nor does its poor environmental record make it an outlier among the other emirates. The UAE's overall per capita consumption of fossil-fueled energy is among the world's highest, and, in a nation drenched in sunshine, only 7 percent of its domestic power comes from renewable energy. Since it has very little groundwater, most of its potable water supply is delivered through seventy power-hungry desalination plants, accounting for 14 percent of the world's total production of desalinated water.[21] If the Emirates, with their jumbo

20 Andrew Ross, "Saadiyat Island," *Artforum* (October 2015), https://www.artforum.com.

21 Robert Mogielnicki, "Water Worries: The Future of Desalination in the UAE," Arab Gulf States Institute in Washington, March 4, 2020, https://agsiw.org.

carbon footprint, can someday become green, then anyone can. As long as they have resources, like the UAE's, to do it.

In advance of Dubai Expo 2020, the state's rulers announced that the UAE would achieve carbon neutrality by 2050.[22] Dubai had put itself on the world map by shamelessly appealing to the appetite for luxury gratification, as exemplified by its annual month-long Shopping Festival. With the world's attention on COP28, the emirate would be an unlikely showcase for solving the problems associated with unbridled consumerism. Unlike in the multiple world-record-holding Dubai Mall, where, even at the height of the summit, I did not find any climate messaging, the Expo grounds were festooned with banners promoting the 2050 net-zero goal, and there was a palpable buzz about the sizable state funds that were being earmarked for solar farms and other forms of clean energy. In that same vein, COP28 was meticulously curated as an opportunity for the UAE's rebrand as a bold leader in the business of green solutions.

Much of my time at the climate summit was spent in conversation with exhibitors inside the Green Zone's big pavilions—with names like Energy Transition Hub Knowledge Hub, Greening Education Hub, and the Climate Finance Hub. A Climate Justice pavilion had made an appearance at COP27 in Sharm El Sheikh, but it was not approved for the Dubai meetings. The lavishly constructed UAE booths took pride of place, illustrating how the new net-zero goals would be achieved with innovative technologies that harvest renewable energy, minimize waste, and perform biomimicry. In the exhibition halls, there was no mention of oil itself—unarguably the elephant in every room. The exhibits were manned by Emirati employees, in freshly laundered *dishdasha*, the long white robe favored by Gulf men, or the black *abaya* and *shaila*, worn by women. This conspicuous public presence is uncommon in the UAE, where nationals, alienated by the sheer density of guest workers, have increasingly withdrawn to their walled compounds and are rarely visible in most urban environments that are not heavily air-conditioned. They were here, as employees of state-backed companies, to advertise the wares

---

22 "The UAE and the 'Net Zero by 2050 Strategic Initiative': Objectives and Challenges," Emirates Policy Center, October 20, 2021, https://epc.ae/en/.

and to do business with prospective buyers. One of the most popular exhibits was hosted by Masdar, the country's well-funded renewable energy company, chaired by Al Jaber himself, which exports and installs technologies in the MENA region, sub-Saharan Africa, and Asia. The models on display were surrounded with inspirational messaging like "Be the Impact Maker," "Treat Carbon Like Money," "We Make Amazing Happen," or "The Best Investment is Planet Earth," and the air was thick with the gaseous banter of tech utopianism. Leaving aside the gee-whizzery that accompanied many of the magical "solutions" on offer in this marketplace, there were some large-scale projects with clear value.

Knowing that water security is one the region's biggest challenges, I was drawn to the exhibits that promised more energy-efficient ways of desalinating water from the Gulf—already the world's hottest sea. As climate change intensifies, the UAE experiences longer droughts and only receives 10 inches of rain annually, though it is increasingly vulnerable to irregular inundations, as occurred after a mammoth downpour in April 2024 dumped 5.5 inches in one day. Its limited and rapidly depleting groundwater supply is exclusively used for irrigation. Because its traditional desalination plants use flash distillation to heat, evaporate, and condense seawater, they use oodles of energy, and, for every liter of potable water (the UAE processes eight billion bottles a day and exports many of them to thirsty customers elsewhere in the Gulf), the plants produce 1.5 liters of concentrated brine waste to be discharged into the sea. With its shallow waters—typically only a hundred meters from seafloor to surface—the Gulf's salinity is intensified by this industrial process, with disastrous consequences for marine life. By some estimates, salinity levels have increased by 50 percent over the last three decades of water purification. It's easy to imagine that the prospect of being able to float without any body motion, as in the Dead Sea, may be marketed as a tourist attraction in the years to come. More ominous, however, is the threat of an oil spill or algal bloom, either of which will compromise the water supply for hundreds of millions of people in the Arabian Peninsula.

The Dubai Energy and Water Department (DEWA) model on display at COP28 was solar-powered and employed the latest desalination technology of reverse osmosis, whereby high pressure forces sea water through a semipermeable membrane, leaving salts behind. The engineer in charge of the exhibit assured me that it was only a prototype and was some years off implementation. He was vague about the prospect of extracting minerals from the waste, though the capacity to do so could recover commercially valuable quantities of lithium, as well as sodium chloride. Sheepishly acknowledging that the brine waste would still end up in the sea, he assured me that discharges at greater depths would better protect marine species. Besides, DEWA had its own nano-satellites to monitor salinity levels. I told him I had just come from the Saudi pavilion—the extravagant House of Sustainability—which boasted a similar exhibit, but with the addition of wind turbines to power the plant at nighttime. The Emirati engineer was skeptical about the Saudi exhibitor's claim that the entire system could operate as a closed-loop energy process, but he said he would go over and check it out. As for the other looming water threat, the projected sea level rise in the Gulf, no one I asked had any idea how the UAE intends to protect its lucrative shoreline of beaches, not to mention the landmark, manmade islands like the Palm, Bluewaters, and the World, all created for luxury real estate ventures. No matter how much progress on climate action is made at future COPs, there is no future on Earth in which the urban centers of Dubai and Abu Dhabi will not have to relocate much further inland within the next century.[23]

Technologies for transforming some of the UAE's seawater into a resilient supply source have obvious applications for countries in the region; seven out of the ten most water-stressed nations in the world are in the Middle East and North Africa. So, too, many of the other "transitional" energy sources on display, such as green hydrogen, produced with clean energy from water by electrolysis, were promising pathways, forged out of the necessity of moving away from fossil fuels. The Dubai Police, proclaiming to be the first carbon-neutral force in the world, paraded its own hydrogen-powered cruiser.

There was just one problem with all these displays of clean technology. There's no evidence that the UAE—or any other of the

23 Owen Mulhern, "Sea Level Rise Projection Map – Abu Dhabi," *Earth.org*, August 4, 2020, https://earth.org.

world's petrostates—is actually moving away from the extraction
of oil and gas. Just before the first Saudi Green Initiative Forum in
October 2021, energy minister Abdulaziz bin Salman vowed that
"every molecule of hydrocarbon will come out," while announcing
plans to increase the kingdom's oil production capacity by more than
8 percent by 2027.[24] The following year, Saudi Aramco, the world's
second largest company by revenue (next to Walmart), saw a larger
investment in its oil field expansion than any other petroleum firm.
In 2022, on the back of higher prices generated by the Russia-Ukraine
war, the company recorded the highest profits ($161 billion) of any
company in history—more than Shell, BP, ExxonMobil, and Chevron
put together.[25] As the oil boom continues to yield lucre, none of the
Gulf's other national oil companies are reducing their production
capacity.[26] Far from "keeping the oil in the soil" (or sand), as climate
justice advocates insist that we must do, these producers are intent on
extracting and selling every last barrel. They just won't be using so
much of it domestically, where investments in renewables are being
prioritized. The post-petroleum future will flourish at home while
the profits roll in from oil exports to other countries. That was the
Janus-faced reality underpinning the exhibits at COP28. As Adam
Hanieh puts it:

> The Gulf states see no contradiction between an embrace of
> "low-carbon solutions" and pursuing the path of accelerating
> fossil fuel production. Importantly, however, this is not just a
> rhetorical exercise in greenwashing: to a significant degree, the
> expansion of the renewable sector is a necessary step towards
> enabling the Gulf states to sell more oil and gas. With very high
> levels of energy consumption at home, the domestic substitu-
> tion of oil and gas with alternative energy sources means that
> more fossil fuels can be made available for export. Indeed, such

---

24 Javier Blas, "The Saudi Prince of Oil Prices Vows to Drill 'Every Last Molecule,'" *Bloomberg*, July 21, 2021, https://www.bloomberg.com.
25 Summer Said, "Saudi Aramco Posts Record $161 Billion Profit for 2022,"
Wall Street Journal, March 12, 2023, https://www.wsj.com.
26 See the special issue of *Middle East Report* on "Post-Fossil Politics" (Summer
2024), edited by Shana Marshall, Laleh Khalili, Kendra Kintzi and Deen Sharp,
https://merip.org.

reasoning is explicitly behind Saudi Arabia's plan to generate half of the country's electricity from renewables by 2030 (which would be faster than most other parts of the world, including the European Union). As Prince Abdulaziz bin Salman put it, such a shift to renewables is envisioned as a 'triple-win situation': greater oil exports, cheaper energy bills at home and the prestige of meeting emissions targets.[27]

The Gulf's petro states are pursuing this triple-win strategy because they can afford the domestic transition, unlike less developed countries without the resources to install renewable technologies along with the low-carbon infrastructure required to distribute clean power. Given the paltry amounts of "climate financing" allocated by Global North countries, poorer nations have no choice but to continue importing oil from the Gulf's petro giants. However, Hanieh also points to an emerging "East-East hydrocarbon axis linking the oil and gas reserves of the Middle East with the production networks of China and Asia." The mercurial growth of Asian commodity production has been largely powered with Gulf oil, and a series of East-East trade agreements have reinforced the interdependent interests of national ruling classes and corporate elites in both regions. The vast majority of Gulf oil exports go to Asian countries. What's more, most of the massive outflow of petrodollars from this trade does not circulate, and is not captured, in Western markets. Its regional absorption by what Hanieh calls "*khaleeji capitalism*" has elevated the GCC member states to a more dominant global investor role, not only in energy but also other key economic sectors.[28] If fossil capitalism is to be dealt a death blow, it will have to occur outside of the old metropole. COP28, as I would learn, was not a historic break from the rule of hydrocarbons, but it was a milestone in the move away from the West's hegemonic sway over decisions about the future of energy.

---

27 Adam Hanieh, "A Transition to Where? The Gulf Arab States and the New 'East-East' Axis of World Oil," in *Dismantling Green Colonialism Energy and Climate Justice in the Arab Region*, ed. Hamza Hamouchene and Katie Sandwell (London: Pluto Press, 2023).

28 Adam Hanieh, *Capitalism and Class in the Gulf Arab States* (New York: Palgrave Macmillan, 2011).

In the meantime, we will expect to see the Gulf states use their petrodollars not only for their own infrastructural development but also to further their image-building by competing with each other's showcase giga-projects. Who, for example, can assemble the most sustainable utopian city? After Abu Dhabi built Masdar, allegedly the first zero-carbon, car-free city, with a population at buildout of 15,000, the Saudis responded with the fantastical blueprint for NEOM (or THE LINE) a 170-kilometer linear city, 200 meters wide and 500 meters high, stretching without interruption or variation from the Red Sea into the desert. Entirely powered by solar and wind energy, it was projected to house as many as nine million residents in a variety of avant-garde architectural habitats. Not surprisingly, the human cost of this scale of construction has been swept under the magic-carpet hype surrounding the project. True to the repressive reputation of Crown Prince Mohammed bin Salman, NEOM's initial building phase resulted in the forcible eviction of Bedouin communities from the Howeitat tribe, and reports of grave human rights violations against construction workers on the site generated calls to boycott the project.[29] According to *Kingdom Uncovered: Inside Saudi Arabia*, a 2024 BBC documentary by an undercover journalist, 21,000 migrant workers had died laying the foundations for NEOM, and another 100,000 had gone missing.

THE UNFRIENDLY STATE OF LABOR

Every building in Dubai's attention-grabbing skyline is mired in a story of labor exploitation, even those required, under the new net-zero policies, to earn LEED certification for green construction. The UAE is hardly the only country hosting extravagant construction projects deemed to be environmentally sustainable while their architects, engineers, and project managers turn a blind eye to the well-being and rights of those who are actually doing the building. The ethical codes of architectural associations are strong on green principles, but are notoriously silent on responsibilities for worker protection

---

29 Ruth Michaelson, "'It's Being Built on Our Blood': The True Cost of Saudi Arabia's $500bn Megacity," *The Guardian*, May 4, 2020, https://www. theguardian.com.

and welfare.[30] When asked, in 2014, about the deaths of construction workers engaged in building her iconic Al Wakrah World Cup stadium in Qatar, Zaha Hadid, the starchitect of some of the Gulf's most mind-blowing structures, infamously observed: "I have nothing to do with the workers," because "it's not my duty as an architect to look at it."[31]

Strictly speaking, Hadid was right; it is the host government's job to set and enforce labor standards. But her comments were taken badly because they highlighted how prominent creatives who lend their names to the brand-building enterprises of illiberal states are insulated from concerns about, and responsibilities for, basic human rights. The same principle of non-liability applies to the business of greenwashing on the part of corporations and government entities. The Gulf states are hardly alone in this racket, but they provide the best evidence that getting to net zero purely through techno fixes—such as renewable technologies or waste elimination—is no guarantee of ecological justice for any society, let alone ones ruled by highly repressive royal families. A petrostate with an autocratic monarchy and an overwhelming majority population of indentured workers is arguably the worst place to look to for guidance on social sustainability. In theory, city-states like Dubai and Abu Dhabi are in a well-resourced position to provide models of green urbanism that we desperately need, but their development as fundamentally undemocratic, *herrenvolk* polities disqualify them from consideration on all but technocratic grounds.

The ethnic supremacy of native-born Emirati, who compose less than 12 percent of the population, depends on their will to keep their families "pure" from any foreign blood. If they do not marry an outsider, they are well looked after. The country's nationals are beneficiaries of one of the world's most bounteous welfare states, with free education and healthcare, zero-interest land grants and giveaways, generous marriage dowries, free or subsidized housing,

---

30 See Peggy Deamer (ed.), *The Architect as Worker: Immaterial Labor, the Creative Class, and the Politics of Design* (London and New York: Bloomsbury Academic, 2015).

31 Vanessa Quirk, "Zaha Hadid on Worker Deaths in Qatar: 'It's Not My Duty as an Architect,'" *Architecture Daily*, February 26, 2014, https://www.archdaily.com.

and a range of soft loans and financial support for other needs. Tribal monarchies have survived in the Gulf by allocating these benefits in return for the loyalty and political quiescence of their citizenry. The generosity of the royal families is attendant on the near-total absence of political and media freedoms. This arrangement has its roots in tribal patrimonialism, but it was encouraged by British authorities in response to the threats of Nasserite secular nationalism and the Omani communist insurgency in the 1960s, and it has served the Gulf monarchies well to this day; the political stability of their rule and tight social control is good for the international business they court.

On the face of it, a visitor to decadent Dubai would see all of the trappings of an ultra-market civilization; "a society," as Mike Davis put it, "that might have been designed by the Economics Department of the University of Chicago," and one that has "achieved what American reactionaries only dream of—an oasis of free enterprise without income taxes, trade unions or opposition parties."[32] Yet, lavish welfare benefits are not what we associate with neoliberal economic policies. These handsome allocations are the patrimonial price to pay for the citizenry's consent to be ruled through absolute authority.

Such privileges, financially secured by Abu Dhabi's treasure chest of petroleum revenue, are also what sets Emirati nationals apart from the vast servant class that attends to their daily needs, builds and maintains the urban infrastructures, and provides every kind of service, including the largest sex tourism sector in the Middle East. Dubai brags about its cosmopolitan population, drawn from as many as two hundred ethnic groups, but almost none of them have any permanent residential rights, or any enduring right of belonging to the city where they make their lives as well as their livelihoods.[33] This multi-class army of foreign nationals includes well-educated expat professionals as well as unskilled laborers, but they can all be deported at a moment's notice. The *kafala* labor system requires each of them to have a sponsor, a potentially profitable role that can be filled by any Emirati individual, whereas the recruitment trail, in Anjali Kamat's

---

32 Mike Davis, "Fear and Money in Dubai," *New Left Review* 41 (September/October 2006), https://newleftreview.org.
33 Syed Ali, *Dubai: Gilded Cage* (New Haven: Yale University Press, 2010).

words, involves "a vast underground network of tens of thousands of recruitment agents, brokers, and sub-agents peddling the fantasy of the Gulf as the destination of choice."[34]

This system is set up for maximum exploitation and formal disenfranchisement. Debt plays a key role in this arrangement. Most of the expatriates are working in conditions of bonded labor; they would not be toiling in substandard conditions if they did not have recruitment debts to pay off, and they are constructing or maintaining buildings that are themselves heavily debt-leveraged. Anyone serious about ecological resilience would give low marks to this kind of social economy, fueled by dicey credit and serviced through indenture.

Most rich countries have a badly treated, and heavily indebted, workforce of migrant workers, but they are typically a small portion of the population. The overconcentration of this highly precarious labor pool in the UAE and other Gulf states is a difference in kind, not a difference in degree, resulting in deep structural inequalities that amount to a form of apartheid. The "separateness" is economic, primarily, between the native-born and the immigrant classes who are further sorted by job segregation and ethnic discrimination; an occupational hierarchy allocates jobs, in descending order, to Westerners and white South Africans, Arabs, Sub-Saharan Africans, and South and East Asians. But the overriding partition is political in nature, secured by the state's buffering of national citizenship for a small minority of the population. There are courts of appeal but no significant degree of due process for the nonnationals who make up the vast majority, and very few laws that adequately protect their well-being.

Labor rights abuse is what drew me to the UAE in the first place. In 2006, Human Rights Watch issued a scathing report (*Building Towers, Cheating Workers*) on the substandard conditions faced by migrant workers employed in the construction boom.[35] Under the

---

34 Anjali Kamat, "The Men in the Middle," *Dissent* (Spring 2015), https://www. dissentmagazine.org. See also Paula Chakravartty and Nitasha Dhillon, "Justice for South Asian Migrant Workers in the Gulf Requires a New Kind of Solidarity," *In These Times*, February 10, 2016, https://inthesetimes.com.

35 Hadi Ghaemi, *Building Towers, Cheating Workers: Exploitation of Migrant Construction Workers in Dubai* (New York: Human Rights Watch, 2006).

kafala system of worker recruitment, workers arrive, mostly from South Asian countries, heavily indebted and tied to their sponsor or employer on short-term contracts. With few labor protections, they are subject to harsh work discipline and withheld passports, and are beaten and deported if they complain or go on strike. Wage theft is rife, as is the provision of excessive work hours. The *kafala* regime has colonial roots in British labor indenture, but the labor camps, total surveillance, and hazardous working conditions, often in extreme heat, are the institutional engine of the UAE's growth.[36]

The year after the HRW report, NYU, my employer, announced that it would build an overseas campus on Saadiyat Island. I contacted Sarah Leah Whitson, director of the HRW's MENA division, to consult about how to push NYU to adopt fair labor standards, and, along with some colleagues, cofounded the NYU Fair Labor Coalition to carry the initiative forward. As a result of our pressure from within, NYU ended up adopting a strong code of conduct for contractors engaged in building the campus, but was not able to persuade its Abu Dhabi paymasters to hire an independent labor monitor to oversee the implementation of the regulations. The monitor selected for the task was heavily beholden to the state and failed to do its job. Consequently, wages were underpaid, and a two-day strike, involving 3,000 workers, was brutally repressed with arrests, beatings, and deportations. Media exposure of these code violations, on the front page of newspapers like *The New York Times* and *The Independent*,[37] was a sharp blow to the reputation of the university as well as to the Abu Dhabi authorities who had courted and bankrolled this outpost

---

36 Natasha Iskander provides a useful historical analysis of the *kafala* system in *Does Skill Make Us Human?* (Princeton: Princeton University Press, 2021). See also, Adam Hanieh, "States of Exclusion: Migrant Work in the Gulf Arab States, in *Just Work?: Migrant Workers' Struggles Today*, ed. Aziz Choudry and Mondli Hlatshwayo (London: Pluto Press, 2016), 41–60; Robert Vitalis, *America's Kingdom: Mythmaking on the Saudi Oil Frontier* (Stanford: Stanford University Press, 2006).
37 Ariel Kaminer and Sean O'Driscoll, "Workers at N.Y.U.'s Abu Dhabi Site Faced Harsh Conditions," *The New York Times*, May 18, 2014, https://www.nytimes.com; Tom South and Archie Bland, "Dubai Workers for British Firm Beaten by Police Over Strike," *The Independent*, May 24, 2014, https://www.independent.co.uk/us.

of a liberal institution to demonstrate that academic freedom could flourish in a society known for its illiberalism.

In the meantime, the Guggenheim Museum announced it would also be licensing a branch on Saadiyat Island. Armed with lessons learned from the NYU campaign, in 2011, I helped establish the Gulf Labor Artists Coalition, an international group of artists and critics, to put pressure on the museum. In both instances, our strategy was to "leverage the brand"—a formula devised by the anti-sweatshop movement in the 1990s in its efforts to tarnish the brands of producers, like Nike and the Gap, with the taint of human rights abuses. Unlike garment factories, which can be moved overnight if their practices are put under scrutiny, these museums and universities are fixed in place, and as liberal institutions, they have built-in ethical values that can more easily be targeted than corporations. The guiding principle of both coalitions was that no one should be asked to teach, study, curate, or exhibit their work in classrooms or galleries built on the backs of labor abuse.

Gulf Labor used a variety of tactics in our Guggenheim campaign: an international boycott of the Abu Dhabi branch, signed by more than a thousand artists, critics, and curators; several occupations of museum branches (through our direct-action arm G.U.L.F.–the Global Ultra Luxury Faction) in New York and Venice; agitprop artwork (through our *52 Weeks* campaign); and direct negotiations with museum officials. As artists and writers, Gulf Labor members were able to say and do "creative" things that more sober organizations like Amnesty and HRW could not, and for a while, after these NGOs were banned, we were the only group able to visit Abu Dhabi labor camps to take testimony from workers themselves. After the Arab Spring, the UAE intensified its crackdown on dissent, resulting in the jailing of dozens of pro-democracy advocates. As we stepped up our investigative fieldwork in Abu Dhabi, a series of travel bans ("for security reasons") fell into place. In addition to myself, the artists Walid Raad and Ashok Sukumaran, and the critic Guy Mannes-Abbot, were all prevented from entering the country.

The embargo did not surprise us—we were closely tailed during our last visit to Abu Dhabi—and the resulting media coverage helped to publicize the campaign. Professional organizations in the academy and the art world (including the American Association of University

Professors and the International Committee for Museums and Collections of Modern Art) issued statements of support. Leading museum directors and curators condemned the bans, while administrators at NYU and the Guggenheim were left to contemplate how and why their Emirati partners had allowed them to be embarrassed in such a public manner. Cynics who had opined, from the outset, that Saadiyat's cultural district was simply window dressing for an authoritarian society took some pleasure in seeing their views vindicated. But for Gulf Labor, the ban was a huge setback. Our campaign aspired to be a solidarity initiative, and so the restricted access hampered our capacity to build relationships with Gulf migrant workers. Undaunted, we pursued a dialogue with the Guggenheim trustees about how to realize the three demands of our campaign: a debt-settlement fund to reimburse workers for their heavy recruitment fees, establishment of a living wage, and the right to worker representation. In the spring of 2015, we brought HRW and some big labor guns, including the International Trade Union Confederation (ITUC) and the Building and Woodworkers' International (BWI), to a summit meeting with museum officials. Shortly thereafter, the museum broke off all contact with us.[38] The Guggenheim Abu Dhabi was stalled for several years as a result, and only recommenced construction in 2022, safer now from prying eyes.

In the interim, several of the other Gulf Cooperation Council (GCC) countries—Bahrain, Kuwait, Oman, Qatar, Saudi Arabia— that employed the kafala system passed a series of reforms. Under intense international criticism for the thousands of worker fatalities in stadiums built for the 2022 World Cup,[39] Qatar declared that it had abolished its kafala system, but the changes have been slow, and limited, in many cases, to an increase in labor mobility—workers are now allowed to change their employer at the end of their contract. At COP28, I ran into Sharon Burrow, former General Secretary of the ITUC, who acknowledged the reforms, in Saudi Arabia and

---

38 Colin Moynihan, "Labor Protesters to Resume Guggenheim Demonstrations," *The New York Times*, April 17, 2016, https://www.nytimes.com.
39 Pete Pattison et al., "Revealed: 6,500 Migrant Workers Have Died in Qatar Since World Cup Awarded," *The Guardian*, February 23, 2021, https://www.theguardian.com.

Qatar in particular, but lamented that hardly any progress had been made by the UAE.

All of the COP28 staff were guest workers themselves, and, among other things, they were responsible for policing the micro-conduct of participants. On several occasions in Dubai when I was breaking the rules—sitting in a restricted area or wearing the wrong clothing to my hotel gym—I was reproached by the attendants in question. "Please obey the regulations," I was told, "or I will lose my visa and get deported." The appeal was earnest enough and did not appear to be a scripted comment. It certainly worked on me, but it was also a point of entry into conversations about the current state of kafala. One of my interlocutors, from the Indian state of Kerala, insisted that he could change his employer anytime, while another told me that he could only do so on renewal of his three-year visa. Others were not sure at all, nor did they know how to find out.

No one I questioned felt that their wages were adequate compensation for their work, and they all acknowledged their recruitment debts. "We make more money than we could at home," a Nepalese security guard admitted, "but we also owe a lot of money." In his case, he had put up his family farm as collateral for a loan to secure passage to the UAE. His work buddy, who was directing pedestrians away from COP28's electric scooter lanes, explained that his own small landholdings were worthless because of the severity of droughts in his foothill province. He had borrowed from relatives as well as the recruitment agent, and so many people in his village were relying on remittances from his shot at the "Gulf Dream." Everyone I spoke to had plans to move on from the UAE. In this matter, they had little choice, since their transience was more or less mandatory until recently. Under an amendment to the residency law in 2021, expatriates can now apply for a five-year post-retirement visa if they have worked in the UAE for fifteen years, are fifty-five or older, and can provide proof of financial stability. By that point, they will have contributed immeasurably more to Dubai than its nationals, who are often portrayed as idlers, overly reliant on their lavish state welfare benefits and their unearned income from rents. A middle-aged taxi driver who hailed from Pakistan pointed out that he "knew the city and its people better than anyone who is a citizen," and that "although my own children were born here, they will probably not be able to stay, and I have to pay for them to go to school and to see doctors." "Where will they go," he asked, "if I die or lose my visa?"

As climate change impacts kick in, the UAE's guest workers are increasingly recruited from climate-stressed regions of the world, and they can be sent back during an economic crunch.[40] Although it is not easy to distinguish between economic migrants and climate refugees, those who are primarily set in motion by climate change already number in the tens of millions. Heat stress, drought, water inundation, and other sudden-onset climate disasters force them from their homes and rural lots to seek employment abroad where they routinely face job insecurity, discrimination, and abuse in labor-intensive industries like construction. To compound the injury, the money they send back home is often needed to reconstruct their damaged houses and lots, so they end up being out of pocket.

Such migrants often find themselves working outdoors, in locations like the Gulf, where unbearable summer heat stress is not only a direct threat to their health but is itself a side effect of the carbon-intensive infrastructure they are employed to build. In September 2023, as the Dubai Expo site was being readied for the COP meetings, FairSquare reported that migrant construction workers, on two separate days and at three different sites, were put to work in extreme heat. This order was in clear violation of the UAE's laws designed to protect outdoor workers from heat stress between 12:30 p.m. and 3 p.m. in summertime months, when the temperature in some areas can reach up to 50°C (or 122°F), amplified by hot winds and high humidity. "Of course, I get headaches and feel dizzy," said one worker, "everyone in this heat does. This weather isn't for humans, I think."[41] In November 2024, another human rights and labor organization, Equidem, released a report showing that migrant workers across the Gulf region were increasingly being forced to work in wet-bulb temperatures of more than 45°C; the internationally agreed-upon

---

40 In *Does Skill Make Us Human?*, Iskander reports that Gulf labor recruiters increasingly target workers who are literate and skilled but who lack opportunities because of their climate-stressed locations.
41 "This Weather Isn't for Humans," FairSquare, London, October 2023, https://fairsq.org.

recommendation of the ILO is that working beyond 32.1°C carries extreme health risks.[42]

Pell-mell growth in the Gulf's cities has already taken a high toll on workers who die of heatstroke or cardiovascular failures while on the job. In the Gulf heat, body temperatures rise too high for organs to function properly, causing respiratory, heart, and kidney failures. Even in December, working outdoors in midday temperatures between 80 and 90 degrees was a challenge. Two men from Bangladesh who were mixing cement in a residential lot wore head scarves soaked in water. I asked whether their employer gave them the mandatory midday break during the hotter months. "Sometimes," said one, "when the work is slower." But "if we are on a job that needs to get done, we work all the time." I asked, "Isn't that against the law?" His mate shrugged, and said, "The boss will just pay the fine, he can afford it." Under these circumstances, they had no choice, even though their health was at great risk. "There is no one to complain to," he added.

Migrants are also the preferred workforce for emergency labor missions in the service of disaster capitalism. In the US, they are regularly recruited in the wake of hurricanes and wildfires by predatory disaster-recovery firms primed to profit from calamities.[43] The cleanup and rebuilding work is dangerous, often involving toxic sludge and other hazardous materials, and observance of occupational safety standards ranges from low to nonexistent. Displaced from their own homes in "frontline communities," these migrants are now frontline workers in high-emitting countries that have the resources to remedy domestic loss and damage but choose to do so only by exploiting the cheapest and most vulnerable workforce.[44]

The UAE and other GCC states may be outliers in their population demographics; no other country has anywhere near their proportion of guest workers. But as climate change intensifies, so will

---

42 *Climate Change in the World of Work: Violence at the Systemic Edge*, Equidem, November 2024, https://www.equidem.org.

43 Sarah Stillman, "The Migrant Workers Who Follow Climate Disasters," *New Yorker*, November 1, 2021, https://www.newyorker.com.

44 Saket Soni, *The Great Escape: A True Story of Forced Labor and Immigrant Dreams in America* (Chapel Hill: Algonquin Books, 2023).

the displacement and migration of large numbers of people, forced to work abroad in countries that only want their dirt-cheap labor and will exclude them from any of the social benefits that come with permanent residency. Far from an anomaly, Dubai's capitalist oasis may turn out to be a prototype for the eco-apartheid citadels of the future, where "essential work" is performed by rightless migrants who are allowed entry only to occupy informal jobs without any workplace or social protections. In this regard, they already are the "canaries in the coal mine," and so the struggle for their labor rights is of paramount importance. Unions like the Building and Woodworkers International are at the forefront of this movement, but, unlike in the case of notoriously abused workers in the garment or agricultural sectors, the construction trade has no leeway for a consumer campaign to foment public disgust.[45]

UNFCCC's primary goal of mitigating and adapting to climate change is also supposed to offset uneven impacts on workers, communities, and habitats. "Just Transition" is the umbrella term for ensuring that the shift away from nonrenewable energy sources does not replicate existing inequalities or introduce new ones. The term originated in the US in the 1970s, when industrial employers responded to the first wave of strong environmental laws by claiming that the new regulations would force them to lay off workers. Trade unions and environmental groups saw a common cause and were in agreement that the move to hold polluters responsible should not come at the cost of decent jobs. They insisted that any transition to a greener economy should protect livelihoods as well as ecosystems. Energy sector unions have often been on the wrong side of the green aisle, but these early initiatives like Environmentalists for Full Employment (formed in 1975), the Labor Committee for Safe Energy and Full Employment (1980), and the OSHA/Environmental Network (1981) showed that an alternative response from the labor movement was possible. Among the more concrete union propos-

---

45 See Building and Wood Workers International, "Adapting to the Heat: Existing Global Responses for Workers in Construction, Building Materials, Wood, and Forestry Industries" (2024), https://www.bwint.org. In July 2024, a new global campaign to combat heat stress, led by the governments of Brazil and the USA, with the support of the ILO, was launched; see https://www.ilo.org.

als to address job loss in the dirty industries was the Oil, Chemical, and Atomic Workers' president Tony Mazzocchi's idea, in the early 1990s, of a "Superfund for Workers," which would provide four years of retraining at full wages and benefits to workers displaced by environmentally-driven shifts in production—the costs to be borne by employers in the extractive or fossil fuel industries.[46] As for the economic benefits of newly created green jobs, a series of studies, dating from the 1970s to the present day, showed that investment in decentralized and labor-intensive green enterprises would create many more jobs than in the highly automated, capital-intensive and geographically centralized facilities of the fossil energy sector. These estimates punctured the employers' myth, recirculated by some unions, that green industrial policy would only result in job losses.[47]

Government officials blithely promoted green jobs as part of the recovery from the 2008 financial crash, but all too often failed to recognize that fossil-industry jobs were heavily unionized and that the entrepreneurial founders of renewable energy firms wanted to be union-free. It was not until the Biden administration's signature 2022 climate change law, the Inflation Reduction Act (IRA), which obliged recipients of public investment in clean energy to provide well-paying and union jobs, that the principle of "just transition" made its way into state policy. In the wholly private sector, workers needed to win for themselves, as they did with the historic UAW contract, negotiated just before COP28, which ensured that well-paying union jobs would be protected in auto facilities retrofitted for the transition to electric vehicles.[48]

In the meantime, the international labor movement lobbied hard to insert "just transition" into the UNFCCC lexicon of actionable language. For almost two decades, workers' rights were regarded as a

---

46 For these early efforts, see Richard Grossman and Gail Daneker, *Energy, Jobs, and the Economy* (Boston: Alyson Publications, 1979) and Richard Grossman and Richard Kazis, *Fear at Work: Job Blackmail, Labor, and the Environment* (New York: Pilgrim Press, 1982).

47 Andrew Ross, "The Greening of America Revisited: Can the US Create High-Skill Green Jobs?" *New Labor Forum* 19, no. 3 (October 2010): 40–47.

48 Katie Myers, "The UAW Ratifies a Contract," *Grist*, November 20, 2023, https://grist.org.

sideshow to the main stage of climate policymaking. Recognition had to wait until the 2015 Paris Agreement, where the term was included in the preamble. By then, the meaning of just transition had broadened to incorporate other aspects of the climate justice movement, and beyond that, to guarantees of gender, racial and sexual justice. It had become the rallying cry for a new order, in which the energy transition was an opportunity to create a society that repaired past harms and built a more equitable future.[49] COP28 reminded us of the concept's labor roots by instituting UNFCCC's first Just Transition Work Programme, designed to promote labor rights and social protections.

At COP28, I spent an entire day attending the meetings of international trade unionists. The UAE, which only abolished slavery in 1971, was particularly unfriendly terrain for the labor cause. Strikes are quite common, though they are seldom reported, but collective bargaining is illegal and so one of main thrusts of just transition for workers—to protect union jobs in the shift toward renewable energy—was not even applicable. Nor was the thrust from the other direction—to convert the insecure jobs of migrant workers into formal livelihoods—any more feasible in the UAE's empire of precarity. At the trade union meetings, the mood was still defiant but hopes for any breakthroughs in Dubai were running low. Nor was there much enthusiasm for the next climate summit, to be held in Baku, capital of another petrostate, whose oil and gas make up 92 percent of its exports and provide two-thirds of the state budget. Azerbaijan, which began pumping oil as early as the 1840s, before anywhere else on the planet, has an even more repressive, single-autocrat style of government; unions are legal, though their political activities are highly restricted. The more progressive labor factions had their sights set on Brazil's hosting of COP30 in 2025, in the Amazonian city of Belém do Pará, where the presence of strong environmental and labor movements would provide more traction for progress on both fronts.

---

49 Edouard Morena, Dunja Krause, and Dimitris Stevis (eds.), *Just Transitions: Social Justice in the Shift Towards a Low-Carbon World* (London: Pluto Press, 2020).

That said, Brazil appears to be adopting the same antipodal formula as the UAE and other fossil fuel powers. Although its economy is too diverse to be considered a petrostate, Brazil is still the world's ninth largest oil and gas producer. Like the US, UK, Canada, Australia, China, Norway, and the major OPEC states, it plans to expand its exports (to become the world's fourth largest within a decade) in order to help finance a renewable energy transition at home.[50] These are the contradictions of petro-nationalism according to the carbon calculus of climate action. Under the Paris Agreement, when countries tally their greenhouse gas emissions, their exports of oil, gas, and coal are not counted. In UNFCCC's accounting system, these fossil fuels only show up on the scorecards of the nations that purchase and burn them. Brazil is not the only country in the region with these ambitions. Buoyed by new discoveries, Venezuela, Mexico, Argentina, Ecuador, Peru, Trinidad and Tobago, Barbados, and Costa Rica plan to expand their extraction of oil and gas. As developing countries, they have the right, under the UNFCCC system, to exploit their hydrocarbon reserves at the same time as overdeveloped nations are supposed to be cutting back on drilling.

ON BRENT BRAVO: A SCOTTISH STORY

As it happens, I had some history of my own as an oil worker, in a country that once harbored its own petro-nationalist aspirations. My stint on a Scottish oil rig spanned only six months, in the summer and autumn in 1978, but my memories of laboring in an outpost of the fossil fuel industry resurfaced during my time in Dubai, and helped me bridge the gap between the meetings I attended about the dilemmas faced by workers and the top-level diplomatic discussions I knew were happening on behalf of nation-states.

Occupational hazards were common on my offshore rig, one of the first drilling facilities to go online in Shell's Brent oilfield. Three hundred miles northeast of the port of Aberdeen, our 300-meter-high platform, Brent Bravo, was moored to the seabed in 140 feet of frigid

---

50 Max Bearak, "Brazil's Clashing Goals: Protect the Amazon and Pump Lots More Oil," *The New York Times*, March 13, 2024, https://www.nytimes.com.

waters. At the time, it was one of the world's deepest extraction fields. Waves in that part of the North Sea can swell to seventy feet, and much higher during gale-force storms. But the workplace challenges we faced were mostly manmade, arising from a blend of class-bound hierarchy and a single-minded focus on extraction that was kindred in spirit to Herman Melville's *Pequod*.

The discovery of North Sea oil in the early 1970s came as a surprise to geologists; the East Shetland Basin was not supposed to host rock formations conducive to hydrocarbon reservoirs. The finding promised an economic bonanza and also helped fuel a nationalist revival in Scotland. The brazen slogan, "It's Scotland's Oil," signaled rosy hopes of an economically buoyant, independent republic. More immediately, the promise of hefty wage packets in a sunrise industry beckoned. I signed on, fresh from securing a degree at Aberdeen University. To protest the college's investments in apartheid South Africa, my last month as a student was spent in the occupation of my university's administration block. Decades later, I would be helping my own college students occupy buildings to protest fossil fuel investments and the Gazan genocide. But, in the late 1970s, there was more concern about "global cooling" than warming, and the pivotal contribution of the oil industry to climate change, while it was known internally to Exxon's researchers and executives in 1977 (and to the board members of Humble Oil, soon to be an Exxon acquisition, as early as 1957), was being kept from public knowledge.

My first job on Brent Bravo, as a steward, was to clean rooms and launder sheets but I was quickly moved up to the kitchen, when Chris the breakfast chef required an extra hand to cook for the hungry souls coming off their twelve-hour night shifts. A working-class conservative from the English Midlands, Chris liked to talk politics, and he needed a leftie to provoke and rail against. As a late hippie, my culinary tastes leaned toward veggie improv, and he tolerated my subtle efforts to subvert the basic menu to that end even though there were few takers among our diners. On the rig, the manual workforce of roughnecks and roustabouts was drawn from the youthful Scottish proletariat: Northern Englishmen filled the more skilled positions; the administrative class hailed from the south of England; and Americans made up the tool-pusher elite who directed the drilling operations. Not a few were "sea dogs," seasoned from long spells on maritime vessels. Unlike in the adjacent Norwegian sector, women were barred from employment on British rigs. Inevitably, then, the homosocial humor on board the rig feminized the stewards and cooks who made up the service staff; I quickly earned the moniker of "Angie."

The Brent platforms had been operational for less than two years, and so the turnover of personnel was still quite high. Within two months, I was promoted to a position above deck as a roustabout, and it was there, on my first shift topsides, that I encountered the tough-as-nails mentality that presided over the day-to-day business of oil extraction. After a few hours on a crane crew, transferring materials from one side of the platform to another, a stack of oil pipes collapsed in a higgledy-piggledy heap, trapping my leg underneath. The task of untangling the pile with a series of delicate crane lifts felt like a jumbo game of pick-up sticks. Remarkably, my limb was unscathed at the end of it, but Bob, the foreman on my crew, was not one to dwell on my good fortune. "I don't want no blood on my deck," he warned me. These are not words you want to hear from a boss, but he was at pains to present himself as a macho West Texan oilman, setting a leathery example for the greenhorn on his watch. In due time, I would learn that he was a devout evangelical, who would commit the lion's share of his salary to an expedition to locate the remains of the biblical Ark somewhere on the slopes of Mount Ararat in Turkey. He was also a dangerous man to be around, because his lack of physical coordination regularly put his crew in peril. When Bob was wielding a heavy tool, we knew to keep our distance.

As general laborers, roustabouts do everything: maintenance and cleaning of the rig; unloading containers and their contents; and ferrying pipes up to supply the drilling floor. Like most jobs classified as unskilled, they actually require a good deal of know-how and technical dexterity. Staying alert on the long shifts was not easy (at least without amphetamines). In a confined space, traversed by heavy moving machinery, there were many opportunities to get injured. Surprisingly, there were no fatalities during my time on Brent Bravo, though I witnessed many disturbing accidents: broken limbs, chemical burns, head fractures. It was the "wild west" era before unions won a foothold in the North Sea sector; worker safety was not a high priority, and regulations were few and far between. For instance, one of our tasks was supposed to be testing the mechanism for lowering lifeboats to the ocean surface ninety feet below, but I don't recall seeing anyone do it. (As it happens, no lifeboats were successfully launched during the Piper Alpha platform disaster exactly ten years later, which took the lives of 167 men.) With no explicit rules on the job to abide by, what protected us?

As is in most high-risk workplaces, the key to safety was crew members following the unwritten code of looking out for one another. We quickly sized each other up, for reliability, strength, and smarts, since all three qualities would be needed if someone was in a tight spot. Mavericks were kept at bay, but not because they seemed to like it that way; no one wanted a rogue male on the team, least of all when we had one of them, Bob, in charge. Bonding with your fellow workers was a social bonus, but first and foremost, it was the thick glue of mutual aid that would keep you from harm. The managerial mind, fixated on gain, sees this synergy as productivity-boosting "teamwork," but that's not the primary reason for this kind of fraternity.

While cowboys were distrusted, there was more than a patina of glamour in an occupation associated with American swagger. Jackie, an inner-city Glaswegian on my crew, was an Americanophile, and seldom failed to remind us of his high devotion to Johnny Cash. Fore-man Bob shared his taste for country music, and so we got to hear many awkward duets from the two of them. But Jackie also fancied himself as a Cali beach boy. As soon as we got off the rig, he headed to Devon for two weeks of surfing, in preparation, he assured us, for a longer sojourn in Los Angeles. Hector, our coworker from a West Highland village, beat him to it, returning with tales of his Golden State adventures that we all regarded as so many sea yarns.

For my two weeks off, I would retreat to my communal farm cottage in the Aberdeenshire countryside, soon to be encroached on by Houston-like suburbs, jerry-built to house the Texans flowing into the region. Intrigued by my reports about life offshore, some mem-bers of the commune I had acted with in college plays talked about forming a repertory company to tour the rigs and provide in-person entertainment to captive audiences. Nevertheless, without women in the cast, it was difficult to imagine putting on the kind of show we would want; nor would our frank views about the predatory character of oil companies have been guaranteed a warm reception, even among the rank and file.

A delegation of important people was coming to visit Brent Bravo, and some bright spark in the administration decided that the rig could use a lick of paint, even though it was relatively new. My crew was given the first shift. Armed with paint guns and strapped into safety harnesses, we were lowered down from the platform edge with instructions to spray the sides with a yellow overlay. The wind was moving at a steady clip, and it didn't take long to recognize the futility of the exercise. If the paint adhered at all, it was in isolated, messy patches. Despite the evidence of our ineffective efforts, Bob, in thrall to his superior, sent us back down, day after day. Being dangled above the ocean and buffeted by a strong wind was no fun, but it was compounded by having zero prospect, in that position, of doing a satisfactory job. At the end of my two-week tour, I handed in my resignation, along with some salty suggestions about improving workplace safety and putting an end to the routine practice of throwing all forms of waste, no matter how toxic, into the sea.

Brent Bravo turned out to be the last in a long series of manual jobs I had held in my native country. In the following weeks, I resolved to take off for a stint in an Israeli kibbutz—a workplace that seemed, on the face of it, to be diametrically opposed, in organization and ethos, to the command capitalism of the oil rig, but had its own role to play, as I would soon learn, in the story of settler colonialism. By the time I returned home, many months later, the advocates for Scotland's first bid for political devolution had been defeated, through a referendum engineered by Westminster to fail (a yes vote required 40 percent of the electorate to succeed). In retrospect, some say that the disappointment created a "lost generation," and I could count myself among them, since I left shortly thereafter.

If independence had come in the decades following, Scotland might have suffered from the contradictions of petro-nationalism, or even the paradox of plenty known to less affluent fossil producers as the "resource curse."[51] But with rising popular opposition to

---

51 James Marriott and Terry Macalister, *Crude Britannia: How Oil Shaped a Nation* (London: Pluto Press, 2021).

oil extraction, the most strident nationalist claims over North Sea oil revenue have weakened or are increasingly compromised. The historic claim to the oil was a strong one. Scots law is different from the legal system of England and Wales, and so up to 90 percent of the UK's oil resources are considered to be under Scottish jurisdiction. The exact share of Scottish oil is hypothetical, however, since Scotland is not a sovereign state. In the vibrant public debate over the independence referendum of 2014, the issue of oil extraction loomed large, influenced by widely varied estimates (by a factor of 60 percent) of how much oil and gas remained to be extracted from Scottish waters. But the oil factor was overshadowed by a bitter face-off over what share of UK public debt an independent Scotland would inherit.

This financial liability became a bargaining chip in the public exchange of threats on the part of leading politicians on both sides of the independence issue. It is surely indicative of the state of popular democracy that this singular factor was pushed into such prominence. Several formulae for measuring Scotland's likely debt obligation were proposed, all in response to the overarching question: for a country intent on sovereign statehood, what is a credible, or sustainable, ratio of debt-to-GDP? All other factors of history, cultural distinction, and social logic—each bolstering arguments for national self-determination that have stretched over centuries—paled in significance before this dull metric of our financialized times.[52] The only criterion that seemed to matter in the end was "sovereign risk"—the investment yardstick used by the international banking community to assess whether a country's past, or potential, record of debt service will comfort the bond markets. For a country with a tradition of strong popular support for the public provision of social services, it would be a tragic outcome if bondholders' rights were given priority over the welfare rights of the citizenry, as has happened so often in the case of Global South nations caught in a debt trap.

In that respect, one of the most innovative proposals floated in the lead-up to the 2014 referendum was an "oil for debt swap," whereby all North Sea revenues would be ceded to the UK in exchange

---

52 James Foley, *The Scottish Economy and Nationalism: Constructing Scotland's Imagined Economy* (Abingdon: Routledge, 2023).

for a hefty write-down of the inherited debt.[53] Such an outcome, according to the report by National Institute of Economic and Social Research, would greatly reduce the fiscal risks that came with independence. But the more attractive feature of this proposal, which the report did not consider, was the prospect of escaping the addiction to fossil fuel revenue that has plagued so many other countries in pursuit of self-determination. Even in those states, mostly in Latin America, where petroleum profits are no longer captured by corporations or comprador elites but are used to expand social service funding, there is a political price to be paid for diverging from the ecological pathways laid out by advocates of low-carbon development.

In Scotland's case, North Sea oil may not, after all, have been the key to independence. It was the wind and the waves that had so harried me and my fellow roustabouts. The country is richer in these renewable resources than it ever was in fossil fuel. By 2024, its generation of renewable energy had outstripped total domestic energy demand, and its estimated wind and tidal power were capable of supplying 25 percent of Europe's energy needs. Today's sunrise jobs lie in harnessing that potential, if workers can be assured of a just transition—as the coal miners of an earlier generation were not. If only we had known, before forty-five billion barrels of oil were sucked out of the seabed, that our extractive labors on Brent Bravo and its ilk were not a beginning, but more like the end of Walter Benjamin's calamitous storm of progress blowing in from Paradise.

But the legacy of that hydrocarbon-fueled storm is still to be reckoned with. Notably, in the 2014 debate over national debt obligations, there was no mention of Scotland's share of UK climate debt owed to Global South countries. The country's most famous engineer, James Watt, is often hailed as the father of the Industrial Revolution for his steam engine inventions that transformed the efficiency of mining and manufacturing operations. Advocates of climate debt often date their calculations to Watt's mid-eighteenth-century moment, when carbon emissions began their ominous rise. While the US has the biggest cumulative share of CO_2 emissions

---

53 Monique Ebell and Angus Armstrong, "Scotland's Currency Options," Discussion Paper No. 415 (London: National Institute of Economic and Social Research, 2013), https://www.niesr.ac.uk/publications/.

over this period of time, the UK's share is larger on a per capita basis, and even greater if emissions from its colonial possessions are taken into account.[54] Carbon-rich industrialization was unevenly developed in the era of European imperialism, but even when the emissions occurred within colonies, the generated wealth piled up in the bank accounts of colonizers. What responsibility did the rich countries bear, now that their former colonies are suffering the brunt of the damage from climate change?

DEBT ABOLITION

COP28 was the latest effort to resolve the contentious bargaining over the responsibilities of high-emitting nations to alleviate the burdens of low-carbon nations. The obligation is often couched as the demand to pay off climate debts. It originated in the concept of ecological debt, introduced in the lead-up to Rio's 1992 Earth Summit as a way for Global South countries to escape from the crushing, fiscal debt trap laid by the IMF and Northern financial institutions. How did the South's external debts to foreign creditors compare with the North's liabilities for environmental impacts from early colonization onwards? These impacts range from the plunder of resources from extractive industry, and all of its associated pollution and biodiversity damage, to the loss of populations from slave trade and colonial wars, and biopiracy of genetic resources from plants and agriculture. However, the net damage from all of this pillage is difficult to quantify. A more reliable measurement could be made of historic carbon debt, appraised on the basis of atmospheric emissions estimates. This obligation subsequently came to be known as climate debt, and its repayment is a core principle of the climate justice movement.[55]

---

54 "How Colonial Rule Radically Shifts Historical Responsibility for Climate Change," *Carbon Brief*, November 26, 2023, https://www.carbonbrief.org.

55 NASA climatologist James Hansen crunched the numbers in a well-publicized 2008 letter to Australian prime minister Kevin Rudd. According to his estimate, the historical carbon debt of the United States was 27.5 percent of the total, though the United Kingdom's obligation exceeded that of the United States on a per capita basis ($33,307 and $31,035 per capita, respectively). See Andrew Ross, "Climate Debt Denial," *Dissent* (Summer 2013), https://www.dissentmagazine.org.

According to the moral accounting of its advocates, the usual North-South relationship of creditor-debtor is reversed. In this context, the North is on the hook.

The specific obligations of carbon-rich countries entered the UNFCCC lexicon through the Kyoto Protocol of 1997, which included the concept of "common but differentiated responsibilities" of nations, but climate activists did not fully take up the call for climate debt justice until the Copenhagen summit in 2009. Anxious to avoid any language about reparations, the US lead climate negotiator, Todd Stern, rejected the idea that developed countries should be retroactively responsible for a problem that could not have been predicted at the time. From this perspective, there was a much narrower window for making repayment claims, on the basis of emissions since 1990, say, when a verifiable link between atmospheric CO_2 and climate change was established by the IPCC's first assessment. The size of the claim would still be colossal, since more than half of all cumulative carbon emissions have occurred during that period of time,

In that reparative spirit, a Bolivian proposal, defining climate debt as comprising an "emissions debt" and an "adaptation debt," was submitted to the UNFCCC in 2010. Payment of emissions debt was supposed to take the form of sharp reductions in domestic emissions of the energy-intensive countries to free up space for others to develop their way out of poverty. Poor countries should not have to forego development opportunities in the interest of containing global emissions. Payment of adaptation debt is supposed to compensate for the damage already done as well as absorbing the costs of preventing future impacts. The G77 bloc of Southern countries (comprising 135 developing nations) and the Alliance of Small Island States, in concert with NGOs and Indigenous groups, have pushed hard since then for these debt obligations to be honored. In response, negotiators from the US and other developed nations have strongly opposed any mention of "compensation" in the final drafts of COP agreements, or indeed any reparative language that implied formal liability for repayment of their climate debts.

In a gesture of minimal cooperation, the G20 bloc of Northern countries made a pledge in 2009 to commit as much as $100 billion annually by 2020 to help poorer nations adapt to climate change. It was "the smallest big number they could think of," as a fellow attendee

in Dubai described it to me. Moreover, contributions to this Green
Climate Fund were voluntary, and they were treated as disbursements of "climate aid," in the spirit of charity, and not as an obligation
in the justice tradition. In a similar vein, compliance with emissions
reductions on the part of rich countries has been largely discretion-
ary. UNFCCC declared that the firmer commitments laid out in the
Paris Agreement were legally binding, but offered no enforcement
mechanism of penalties, fees, or sanctions for failure to comply.

It was no surprise, then, that the Green Climate Fund failed to
meet its 2020 target, though its donors claimed to have mobilized
$83.3 billion of climate financing. However, an influential Oxfam
report, issued five months before COP28, showed that the real value
of the donations had been much less—$24.5 billion at most.[56] The
authors' estimate was based on three forms of evasion: a) many countries had simply repurposed their official aid contributions as climate
finance; b) much of the funding was not directed toward climate
adaptation objectives; and c) more than half of the financing had
come in the form of concessional loans and not grants. Given how
climate reparations originated in an effort to reduce or cancel the
external debts of the Global South, it is ironic that so much of it has
been disbursed in the form of loans that may have added to those debt
burdens. France, for example, made 92 percent of its contributions
through loans, with Japan (90 percent) and Spain (88 percent) close
behind. In addition, many of those loans have been made through the
IMF and World Bank, the same multilateral institutions widely held
responsible for the Southern debt trap. True to neoliberal form, they
have imposed antisocial conditions on the loans—shrink your public
sector and contract private enterprise to take on the climate adaptation projects. In effect, the Green Climate Fund had been funneling
public monies into private pockets.

In 2023, donors claimed to have finally met the $100 billion
target, just in time for COP28. But, in the intervening years, climate
change impacts and rising emissions have magnified the scale of climate debt far beyond the 2009 numbers. In the lead-up to COP27 in

---

56 "Climate Finance Shadow Report 2023," *Oxfam International*, May 6, 2023,
https://policy-practice.oxfam.org.

Egypt, a group of economists, including Lord Stern, the establishment authority on climate financing, published a UN-commissioned estimate that developing countries would need as much as $2.4 trillion in annual financing to help transition to a low-carbon footing, adapt their infrastructure, and rescue communities hit by climate disaster.[57]

"UNABATED"?

The first day of COP28 saw the announcement of a Loss and Damage Fund, to help vulnerable nations respond with urgency to climate change impacts. It was lauded as a fresh opportunity for G20 nations to show they were responsive to their climate obligations. Yet a mere $700 million was raised in pledges; the UAE, Germany, Italy, and France topped the donor list with at least $100 million each, while the United States embarrassed itself with a promise of only $17.5 million. The UK promise of $75 million was not even new money—it was subtracted from its existing climate finance pledge. The $700 million on offer paled in comparison with what is needed. One study of loss and damage from impacts in developing countries estimated the annual toll at $400 billion.[58] The 2022 floods in Pakistan alone generated $30 billion in damage. The underachieving "victory" of the new Loss and Damage Fund set the tone for the rest of the proceedings in Dubai.

The UNFCCC mechanism for reaching multilateral agreements is a long, unforgiving grind. The discussions that take place in dozens of committees and working groups are all aimed at influencing the language of the final text. It is a process in which every word and comma is subject to exacting scrutiny and contestation. Every participant nation has to consent to the outcome. As a result, any single

---

57 Independent High-Level Expert Group on Climate Finance, *Finance for Climate Action: Scaling Up Investment for Climate and Development*, Grantham Research Institute, London School of Economics (November 8, 2022); "Global South Needs $2 Trillion a Year to Fight Climate Crisis," *Al Jazeera*, November 8, 2022, https://www.aljazeera.com.
58 Julie-Anne Richards et al., "Standing in Solidarity with Those on the Frontlines of the Climate Crisis," The Loss and Damage Collaboration, November 20, 2023, https://www.lossanddamagecollaboration.org.

party can thwart the will of the majority. Nonetheless, those hoping for a definitive blow to fossil fuel power in Dubai were buoyed by the fact that 130 of the 198 countries in attendance (a supermajority in some quorums) declared their support for language calling for a fossil fuel *phase out.* This plurality included the rich EU countries, though none of them register among the world's leading fossil fuel producers.

But the bitter dispute over the inclusion of phase out in the Dubai agreement was complicated by a debate about the meaning of the adjective "unabated," first used in the Glasgow climate pact of COP26. During those meetings, India, the world's second largest coal producer, had insisted on a "phase down" of coal rather than a phase out. This stand was further parlayed by advocates of unproven technologies like direct air capture and storage language into language in the final agreement about phasing down "unabated coal." The inclusion of this term generated uncertainty about the exact nature of "abated" emissions. If a coal-fired power plant successfully captures only 10 percent of the CO_2 it produced, could its operators still claim that the emissions were abated? The IPCC tried to clarify the issue in its 2022 report: "'unabated fossil fuels' refers to fossil fuels produced and used without interventions that substantially reduce the amount" of greenhouse gasses, "capturing 90% or more CO_2 from power plants, or 50–80% of fugitive methane emissions from energy supply." Yet there was still some ambiguity in that wording, providing a loophole that fossil fuel producers were guaranteed to exploit.[59] In advance of the final COP28 agreement, an Oxfam press release sounded the alarm: "abatement linked to fossil fuel phase out is a gift served on a silver platter to the fossil fuel industry they will gladly seize to dodge any commitments."[60] António Guterres, the UN's proactive secretary general, jumped in to more bluntly summarize one side of this war on words: "Not reduce. Not abate. Phase out."

The pressure campaign failed miserably. A leaked letter from OPEC (dated December 6) confirmed that the major petrostates

59 "Why Defining the 'Phase-Out' of 'Unabated' Fossil Fuels is So Important at COP28," *Carbon Brief*, December 5, 2023, https://www.carbonbrief.org.
60 "Reaction to the Possibility of 'Abatement' Language in the Final Text of COP28," Oxfam International, December 11, 2023, https://www.oxfam.org.

would block any mention of a phase out.[61] Five days later, a prerelease of a draft of the final COP28 resolution confirmed that these nations were likely to carry the day. "This obsequious draft reads as if OPEC dictated it word for word," Al Gore fulminated. "It is even worse than many had feared. It is 'Of the Petrostates, by the Petrostates, and for the Petrostates.'"[62]

The final compromise deal, released two days later, called for "transitioning away from fossil fuels" without clear obligations for parties or hard timelines. It outlined, albeit in the vaguest terms, the need for tripling renewable energy capacity. In the weak phrasing of "contributions" that is typical of nonbinding COP agreements, the pact also recommended "accelerating zero- and low-emission technologies" such as carbon capture and storage, the sequestration strategy that is the darling of the petro powers. "Phasing out" appeared only once in the final text, in equally imprecise language calling for the "phasing out [of] inefficient fossil fuel subsidies that do not address energy poverty or just transitions."[63] The agreement sidestepped the heated debate about the meaning of abatement by simply repeating, and slightly amplifying, the Glasgow formulation in its call for "accelerating efforts towards a phase down of unabated coal." This "acceleration" was to be accomplished by the use of "abatement and removal technologies, such as carbon capture and utilization and storage, particularly in hard-to-abate sectors" though there was no attempt to further define what qualified as abatement.

In other parts of the agreement, "transitional fuels"—including liquefied natural gas (LNG)—were given a blessing. This was a particular gift to the US, by far the world's largest producer, whose LNG exports are skyrocketing, especially to European states looking

---

61 Lisa Friedman, "OPEC Leader Tells Members to Block Any Climate Summit Deal to Curb Fossil Fuels," *The New York Times*, December 8, 2023, https://www.nytimes.com.

62 Olivia Rosane, "COP28 on 'Verge of Complete Failure' as Draft Omits Fossil Fuel Phaseout," *Common Dreams*, December 11, 2023, https://www.commondreams.org.

63 UNFCCC, *Conference of the Parties Serving as the Meeting of the Parties to the Paris Agreement*, November 30 to December 12, 2023, https://unfccc.int.

to replace Russian gas in the wake of the war on Ukraine. While the burning of LNG generates less carbon emissions than coal or oil, its extraction, primarily from fracking, and its distribution are energy-intensive and also release high amounts of methane, by far the most potent of greenhouse gasses. In Dubai, more than thirty countries put their weight behind "clean hydrogen" by launching a Declaration of Intent on the Mutual Recognition of Certification Schemes for Renewable and Low-Carbon Hydrogen and Hydrogen Derivatives. But COP28's endorsement of this alternative as a transitional fuel masked the many environmental risks associated with its production and use, while boosting a new investment sector for slightly less dirty energy.

Last but not least, the language urging developed nations to contribute more climate finance lacked teeth. The COP28 parties decided to punt any detailed agreement on the issue to the next year's meetings in Baku, while agreeing to replace the failed Green Climate Fund with the more jargon-intensive New Collective Qualitative Goal (NCQG), tossing a new acronym into UNFCCC's thick alphabet soup. The goal was to restore faith and trust in the process by responding to the real needs of developing countries instead of plucking numbers, like the original $100 billion, out of thin air. While nothing was settled in Dubai, the battle lines over NCQG were drawn quite clearly. Hoping to head off further claims on their public expenditures, G20 nations were looking to "expand the contributor base" by including high-emitter countries anachronistically designated as "developing." By the time COP29 convened in Baku a year later, there was a growing clamor to add China, India, South Korea, UAE, and Saudi Arabia to the list of donor countries and it provoked a bitter fight. But "expanding the base" was also code for relying more on international private capital, or to use the corporate parlance of an official US response, "unlocking the trillions" needed "to achieve finance at scale."[64] After much rancor, negotiators at COP29 ended up striking a deeply unsatisfying deal whereby rich countries pledged to "reach" $300 billion per year in climate finance by 2035; the final text called

---

64 C. Radhika, "Diverging Views on New Finance Goal," *Dubai Climate News*, December 4, 2023, https://www.twn.my.

on "all actors" to make up the shortfall from the estimated $1.3 trillion needed by developing countries to effectively combat climate change.

Multilateral agencies like the World Bank were expected to step up and play a greater role, not just in delivering the aid, but also in backing loans from private institutions. For the Global South, these moves evoked bitter memories of structural adjustment packages—loans conditional on privatization and deregulation, cuts in public expenditures, elimination of barriers to foreign investors, and promotion of export-led development. Instead of directly acknowledging their climate debts, the North's turn to lending from public sources and from private capital was seen as yet another neocolonial opportunity to retighten the debt trap and further advance economic liberalization. For their part, clean tech companies would surely salivate at the prospect of opening new markets for their products, either through leasing or other profitable arrangements, especially if UNFCCC offered a way to secure their investments against default.

The many loopholes in the final COP28 text and its generally ersatz commitments left Indigenous participants, frontline communities, trade unionists, and climate justice activists fuming. For developing countries in particular, the directive to pursue a deep energy transition without access to adequate resources for doing so was a bitter pill to swallow. Bolivia, traditionally UNFCCC's most strident critic, bluntly described the final document as a manifestation of "carbon colonialism," and pointed a finger at the hypocrisy of the developed nations who talked a lot about the "North Star" of "keeping the 1.5 temperature limit alive, while endorsing the expansion plans of Big Oil.[65]

Undaunted, the UNFCCC managers, along with the G20 nations, declared the outcome to be a historic victory: "The beginning of the end of the fossil fuel era."[66] Much was made of the first

65 Meena Raman and Hilary Kung, "Mixed Reactions to Dubai Outcomes," *Dubai Climate News*, Third World Network Update No. 18, December 15, 2023, https://www.twn.my.

66 UNFCCC, "COP28 Agreement Signals "Beginning of the End" of the Fossil Fuel Era, December 13, 2023, https://unfccc.int; Georgina Rannard, "Landmark Summit Takes Direct Aim at Fossil Fuels," *BBC*, December 13, 2023, https://www.bbc.com.

appearance of the words "fossil fuels" in a COP agreement. But the oil barons left Dubai with smiles on their faces, knowing that nothing in the deal would scare off investors, and that their plans to expand production could proceed apace. In the real world, where their companies toil ceaselessly to open new markets and secure new drilling leases, fossil fuels have been *phasing up* rather than down, entering a new bonanza era. The rise in generated gigawatts of renewable energy has not eaten into the volume of fossil fuel production; it has simply added to the total consumption of energy. Indeed, there was barely time to mourn the slaying of the phase out before Al Jaber himself recommitted to business as usual, announcing that the UAE would continue to be a "responsible, reliable supplier of low-carbon energy . . . at the lowest cost," explaining that his barrels of oil were "lower carbon" because their efficient extraction ensured less leakage than those from other sources.[67]

After the deal was announced, a Saudi Arabian representative bluntly declared that the agreements "do not affect our exports, do not affect our ability to sell."[68] Saudi Arabia has a commanding lead over other oil exporters but the list of countries in that club (which includes developed and developing nations) is a long one, and he was speaking for all of their interests. In the US, the Biden administration lauded its legislative record of domestic climate action despite approving thousands of new drilling leases, including the nation's single largest oil project, the massive Willow oil-drilling venture on Alaska's remote North Slope, described by environmentalists as a "carbon bomb." Under Biden, 20 percent more oil and gas licenses were issued than during Donald Trump's first term. In anticipation

67 Fiona Harvey, "COP28 President Says His Firm Will Keep Investing in Oil," *The Guardian*, December 15, 2023, https://www.theguardian.com.
68 "This Deal Doesn't Affect Our Oil Exports, Says Saudi Arabia," *BBC*, December 11, 2023, https://www.bbc.com. Just two months after COP28, Saudi Arabia's energy minister, Abdulaziz bin Salman, told a conference that the "transition away" from fossil fuels was only one choice on an "a la carte menu" of actions from the summit. Fiona Harvey, "Stop Looking for Loopholes, UN Warns, After Saudi Hints End of Fossil Fuels 'Just One Option,'" *The Guardian*, February 2, 2024, https://www.theguardian.com.

of his reelection, Trump himself was busy making deals with oil executives to give them whatever they wanted in return for campaign funding, and, after his inauguration, under the guise of an "energy emergency," he moved to fast-track hundreds of fossil projects in violation of longstanding requirements to assess their environmental impacts.[69]

A few weeks after the Dubai meetings, it was revealed that Azerbaijan, the COP29 host, planned to increase its fossil fuel production by a third over the next decade; most of that output will come from one of the world's largest gas fields near the Caspian Sea, and it will be allocated for export.[70] In keeping with the two-handed template, the country's ruler has been making the most of his efforts to ramp up Azerbaijan's share of renewable energy projects for domestic consumption. Yet the state-owned oil and gas company, Socar, was set to raise the country's annual gas production from thirty-seven billion cubic meters (bcm) to forty-nine bcm by 2033, and had only committed 3 percent of its capital expenditure to a "green energy division" announced just a few weeks after Azerbaijan was appointed to host COP29.[71] By the time COP29 was underway, despite record growth in renewables, new oil and gas projects had been approved around the world in complete defiance of the "phase down" heralded in Dubai. Diplomatic deliberations in Baku took place within walking distance of thousands of "nodding donkey" oil wells that dot the city and in sight of oil refineries. Azerbaijan's president, Ilham Aliyev, opened the proceedings by declaring that that his country's oil and gas were "a gift of God."

Dubbed the UAE Consensus, the final COP agreement in Dubai had cemented the antipodal formula that UNFCCC has enabled over the years; go green at home and sell every last barrel abroad. Arguably

---

69 Lisa Friedman, "Trump Administration Moves to Fast-Track Hundreds of Fossil Fuel Projects," *The New York Times*, February 19, 2025, https://www.nytimes.com.

70 Jillian Ambrose, "COP29 Host Azerbaijan to Hike Gas Output by a Third Over Next Decade," *The Guardian*, January 8, 2024, https://www.theguardian.com.

71 Damian Carrington, "COP29 Host Azerbaijan Set for Major Fossil Gas Expansion," *The Guardian*, October 23, 2024, https://www.theguardian.com.

the best example of this kabuki theater is Norway, which sits at eight on the list of top oil exporters, behind Saudi Arabia, Russia, Iraq, the United States, Canada, UAE, and Kuwait. Norway exhibits the greatest disparity between its impressive domestic record of reduced emissions, which puts it among the few nations to come close to meeting its Paris targets, and its expansion of oil and gas production for export. The summer before COP28, its government approved the development of nineteen new oil and gas fields, including in previously untouched and fragile areas of the Arctic. Norway's domestic report card gives it bragging rights as a good climate citizen, but its appetite for export-led exploitation tells a different story. Indeed, just four days after the COP28 agreement was gaveled, Oslo inked a massive $50 billion ten-year contract to supply gas to Germany to replace the shortfall from Gazprom's suspension of direct deliveries through the Nord Stream pipeline.

The army of petroleum lobbyists in Dubai succeeded in securing the long-term future of their paymasters. As for the rest of us, it was difficult to avoid the conclusion that we had been cast as extras in a vast theatrical enterprise, carefully scripted to provide the lead actors—the villains of the piece—with the stage set they needed to get what they wanted without giving up too much. The COP script has evolved over the years to include some acknowledgment of the climate debts and responsibilities of the wealthy countries; the "transition" proposed in Dubai is supposed to be "just, equitable, and orderly." Talk is usually cheap, but it is costly when it leads to inaction. In the meantime, the debts will pile up, and the fight for climate justice will continue, away from the hubbub of press conferences, politician photo-ops, and grandstanding from the COP diplomats.

TAKING STOCK

There were many reasons to walk away from Dubai in a dark funk of climate pessimism, but any serious move to halt fossil production was never in the cards. Like the UN itself, the UNFCCC has built-in checks and balances to forestall any *abrupt* shift away from fossil-powered growth, especially one that would facilitate an alternative economic order. However much it is "guided by science," the process has evolved as an arena of convenience for nation states not only to showcase their good green intentions but also to apply diplomatic pressure, both individually and through blocs of allies, to protect their own economic growth.

In the case of the former, this kind of participation can amount to little more than virtue-signaling since the pledges made at these meetings often do not result in follow-through. In the US for example, Republican lawmakers have regularly blocked the allocation of climate funds, and the first and the second Trump administrations withdrew from the Paris Agreement. The shortfalls of the Green Climate Fund and the universal failure of states to reach their emissions reduction targets is proof of the gap between the wholesome promises made on COP's international stage and the messy combat zone of domestic politics, especially in countries where climate denialism is a potent factor, and where far-right wing ideology holds that it is a patriotic obligation to exploit every last drop of its fossil reserves.[72] For capitalism's most profitable corporations, the prospect of abandoning these "stranded assets," and leaving trillions of dollars of revenue in the soil, is simply unthinkable.

As for economic growth, UNFCCC has been curated, over the decades, as a reliable vehicle for "sustainable development"—the preferred label for capitalist growth in the form of market environmentalism. First introduced at the 1992 Earth Summit in Rio summit, the concept of sustainable development, long considered an oxymoron by its critics, is supposed to balance growth with social equity and environmental well-being. At best, it has been taken up by policymakers to apply economic levers to reduce poverty and hunger, and to promote public health, gender equality, education, ecosystem survival, urban resilience, and clean energy. At worst, sustainable development has been utilized as a cynical exercise in greenwashing on the part of corporations looking for cover for their short-term profit orientation.

Nowhere is the duplicity more evident than in the boardrooms of Shell, BP, Chevron, ExxonMobil, and TotalEnergies. After issuing post-Paris climate pledges to move "beyond oil" and reduce their fossil

---

72 For an analysis of how closely tied the rise of white nationalism is to fossil power, see Andreas Malm and the Zetkin Collective, *White Skin, Black Fuel*.

commitments, they largely abandoned these targets after the war on Ukraine ushered in an oil boom.[73] The Big Five were still spending millions of PR dollars on environmental protection promises at the same time as they were expanding production and reporting record profits. According to Moody's, in the first eight months of 2023, this cartel generated a record $613 billion in cash flow and made shareholder payouts of more than $235 billion.[74] If marketing won't do the job, then the corrupt practice of carbon offsets will. In 2022, Blue Carbon, a private Emirati company backed by the Dubai royal family, began to purchase vast swathes of land in Liberia, Zambia, Zimbabwe, Tanzania, and Kenya (the size of Britain). Under the guise of "protecting forests," the company will earn the right to sell carbon credits to firms like Shell or BP that want to continue polluting while canceling out their own emissions.[75] In cases like these, sustainable development is just another name for carbon laundering.[76] The first measure announced at COP29 in Baku, declared to be a breakthrough, was the announcement of new carbon market rules to govern this kind of trading.

Even the most progressive advocates of sustainable development have been unable to shake off its central commitment to pro-growth activity. As a result, the paradigm is often summarized as a recipe for the survival of capitalism, not the planet. Among the alternatives, including those that emphasize *human* over economic development, is the commons-based vision of life in deep harmony with nature embodied in the Andean philosophy of *buen vivir* (or *sumak kawsay*

---

73 In February 2025, BP bowed to its activist shareholders in agreeing to a "fundamental reset" that raised spending on oil and gas by a fifth while cutting expenditures on renewable energy by 70 percent. Stanley Reed, "BP to Focus More on Oil and Gas in Strategy 'Reset,'" *The New York Times*, February 27, 2025, https://www.nytimes.com.

74 "Shifting Energy Landscape Creates Uncertainty for Big Oil. A $613 Billion Windfall Provides Options," *Moody's*, December 13, 2023, https://www.moodys.com.

75 Yinka Ibukun and Natasha White, "Dubai Firm's Africa Ambitions Raises Carbon Colonialism Concerns," *Bloomberg*, November 29, 2023, https://www.bloomberg.com.

76 Adam Hanieh, *Crude Capitalism: Oil, Corporate Power, and the Making of the World Market* (London: Verso, 2024).

in its Quechua formulation). This doctrine lay at the heart of the People's Agreement of Cochabamba, adopted at the World People's Conference on Climate Change and the Rights of Mother Earth in 2008, the year after the Copenhagen COP meetings. Approved by an Indigenous-led gathering of world citizens from more than 140 countries, the Cochabamba Agreement was a far-reaching response to the failure of the UNFCCC process to acknowledge the capitalist roots of the climate emergency and to question the "domination of nature" implicit in the paradigm of growth-driven development.[77] It demanded the "decolonization of the atmosphere," the securing of rights for climate refugees, and the repayment of climate debts. Participants also drafted a Universal Declaration of the Rights of Mother Earth[78] which promoted the concept of the "rights of nature," enshrined in the constitutions of countries like Ecuador and Bolivia.

Champions of the Cochabamba ethos of protecting Pachamama (Mother Earth) have pushed back against left-leaning rulers of Andean states who committed to continuing the colonial patterns of resource extraction, even when the revenue was being funneled directly into public infrastructure and poverty alleviation.[79] Four months before COP28, an Indigenous-led coalition in Ecuador culminated a decade-long grassroots struggle by forcing a national referendum in which a large majority of citizens voted to ban drilling in the Amazonian territory of the Yasuní National Park, one of the of the world's most diverse biospheres.[80] The win—which Ecuador's new president Daniel Noboa has tried to ignore—was a resounding blow to the neo-extractivist doctrine that poorer nations should be free to develop their own economies by exploiting their fossil assets.

---

77 "People's Agreement of Cochabamba" (April 22, 2010), https://www.climateemergencyinstitute.com.

78 "Proposal Universal Declaration of the Rights of Mother Earth," https://pwccc.wordpress.com/programa/.

79 Thea Riofrancos, *Resource Radicals: From Petro-Nationalism to Post-Extractivism in Ecuador* (Durham: Duke University Press, 2020); Nicole Fabricant, "Climate Debt: Who Profits? Who Pays?" *NACLA Report on the Americas*, April 22, 2013, 19–21.

80 Angélica María Bernal and Joshua Holst, "Ecuador Votes to Keep Yasuní Oil in the Ground in Historic Referendum," *NACLA Report on the Americas*, August 23, 2023, https://nacla.org.

The fight to keep the oil in the soil of the Amazon or the fragile ecosystems of Alaska and the Arctic attracts widespread popular support because these are all charismatic locations, much easier to romanticize than the lone and level sands of the Arabian Peninsula. For other, less charmed spots, militant methods of resistance are necessary, drawing on a diversity of tactics—lawsuits against energy companies, blockades of shipments and distribution lines, occupations of public space and other disruptions of daily life, hunger strikes and self-immolation, boycotts, and last but not least, property damage, exemplified by the sabotage of SUVs and the detonation of oil pipelines and executive jets. Regarding the use of violence, Andreas Malm has argued, almost every notable resistance movement has used weaponry to destroy property that is legitimately used to spread oppression.[81] The record shows that oil barons are not swayed by protest in the streets, and they are prepared to spend large fortunes to oppose or coopt efforts at state regulation. They will have to be forced to feel there is a personal price to pay for their own violence; fossil capitalism kills thousands every year.

In the meantime, the newest forms of mobilization are helping to make Big Oil a pariah in public arenas. These have included the inventive actions of Just Stop Oil, Ultima Generazione, Extinction Rebellion, Ende Gelände, Climate Strike, Blockade Australia, Earth Uprisings, Code Pink, Climate Defiance, Fridays for Future, Sunrise, and the hundreds of groups affiliated with the Climate Action Network. These groups have built on a tradition of direct action and civil disobedience pioneered by Greenpeace and the radical environmentalists in Earth First, Earth Liberation Front, and the Earth Liberation Army who espoused tactics variously described as "ecotage" or "monkeywrenching." In Canada and the US respectively, the Indigenous resistance of Idle No More and the Lakota-led stand against the Dakota Access Pipeline at Standing Rock posed a threat to the distribution infrastructure of fossil companies. Actions undertaken as water protectors and land defenders have helped to affirm Indigenous moral leadership of the climate justice movement.

---

81 Andreas Malm, *How to Blow Up a Pipeline: Learning to Fight in a World on Fire* (London: Verso, 2020).

These grassroots campaigns bear comparison with divestment decisions by high-profile institutional investors that have resulted in an estimated $40 trillion sell-off of dirty energy stocks. This massive exercise in ethical investing has not, however, extended to core Wall Street institutions. Since the Paris Agreement, leading bankers who churn out climate PR through their embrace of environmental, social, and governance (ESG) indicators are responsible for funneling trillions of dollars into the expansion of fossil fuel activity and industrial agriculture in the Global South.[82] The double-dealing profiles of JP Morgan, Bank of America, Wells Fargo, Citigroup, BNP Paribas, Barclays, and the Royal Bank of Canada are consistent with the real Dubai Consensus—putting on a pretty face for their domestic clients while getting faraway countries hooked on oil.[83]

The record shows that lawmakers and courts can be pressured to halt approvals for targeted sites or even to accelerate the phase outs of coal power, but, if they are in thrall to dirty energy lobbyists and their Wall Street financiers, they cannot be relied on to rein in fossil capitalism with any degree of urgency. The prospect of international state power has been stunted by UNFCCC's shortcomings, but it is not impossible to imagine how things could be different. One of the most fully articulated visions of global governmental action is Kim Stanley Robinson's influential 2020 novel, *Ministry for the Future.* Leaning more towards the utopian genre of "solarpunk" than the climate fiction known for its fatalism, it stands out as a well-engineered story about the success of a multinational agency tasked with preventing global warming. The ministry in question is established after a historic heatwave kills as many as twenty million in the Indian state of Uttar Pradesh. But the decisive shift in action is triggered by the ministry's "black ops" campaign of eco sabotage; polluters are targeted for assassination, private jets are blown out of the sky, coal-fired power plants are disabled, tankers are sunk offshore, and cattle are injected with mad cow disease to force people off beef consumption. Terrorism championed by progressive, albeit rogue, officials catalyzes

82 Dharna Noor, "Banks Pouring Trillions to Fossil Fuel Expansion in Global South," *The Guardian*, September 4, 2023, https://www.theguardian.com.
83 Alex Simon, "In Dubai, a 'Good Vibes Only' Approach to Climate Change," *The New York Times*, December 8, 2023, https://www.nytimes.com.

their success in halting runaway climate change and reducing atmospheric carbon levels. Science fiction can flip any script, of course, but *The Ministry for the Future* is true to the genre's roots by fleshing out a plausible future extrapolated from the science of the present.

The "future is not written," as Rebecca Solnit reminds us in the *Not Too Late* project, which is aimed at combating the fear, anxiety, and depression experienced by many climate-watchers. After all, campaigns around the world have recorded thousands of victories, large and small, as part of the overall movement to decarbonize the atmosphere.[84] Even though the plunging cost of solar and wind power has resulted in a fall-off of interest from investor speculators, bolstering the case for publicly owned and democratically managed firms, there are many examples of rapid transitions to renewable energy. In a little over fifteen years, Uruguay has almost completely eliminated fossil fuels from its electricity production through building more than fifty windfarms and decarbonizing its energy grid.[85] It is well-placed to do so, because it has no oil or coal of its own, and as the wealthiest South American country, it has the resources to change its energy infrastructure. Other countries, like Costa Rica and Iceland, have achieved almost 100 percent production from renewable electricity sources. When he came to power in 2023, Colombia's left-wing president Gustavo Petro prohibited the award of new hydrocarbon exploration contracts, and smaller producers like Belize and Ireland have also banned offshore drilling.

Countries with proven oil reserves and less income find it much more difficult to abstain from extracting it. In Ecuador, thirteen years before its citizens voted, in a national referendum, to ban drilling in Yasuní National Park, President Rafael Correia tested the world's will for climate justice by issuing a wager—if carbon-rich countries and other donors ponied up enough money to indemnify Ecuador for lost public revenue, then the Ishpingo-Tambococha-Tuputini (ITT) area of the Yasuní would be left alone. A single hectare in that patch of the Amazon contains more tree species than all of the US and Canada

84 Rebecca Solnit and Thelma Young Lutunatabua (eds.), *Not Too Late: Changing the Climate Story from Despair to Possibility* (Chicago: Haymarket, 2023).
85 Sam Meadows, "Uruguay's Green Power Revolution," *The Guardian*, December 28, 2023, https://www.theguardian.com.

combined. Donations came nowhere close to the $3.6 billion goal set by Correa (50 percent of the market value of the park's oil reserves). Only $336 million had been pledged and only $13.3 million had actually been delivered by the time his government abandoned the initiative and began to auction off blocks of the ITT to the state-owned Petroamazonas and Petroecuador, and to Chinese oil subsidiaries.

A coalition, led by the youth group Yasunidos, fought back and won a national referendum in August 2023, in which a decisive majority (almost 60 percent) voted against the extraction of almost a billion barrels of oil in Yasuní National Park. The campaign was the first popular referendum in the world that proposed to leave oil underground. It was undertaken in defense of the rights and lives of the Tagaeri-Taromenane Indigenous peoples who live in isolation in the Ecuadorian Amazon. The campaign appealed to the inclusion of the "rights of nature" in the 2008 Constitution (the first country to do so) which authorizes legal action to protect ecosystems from harm. Fifty years of extraction by transnational oil companies (and widespread despoliation by Chevron in particular) had shown the Ecuadorian people that the formula of overcoming poverty through oil revenue was bankrupt. The populations living closest to extraction sites were the poorest and the most poisoned by spills and pollution.

By the time of the referendum, 85 percent of Ecuador's energy supply came from renewables, mostly hydroelectric, while the country's equatorial location guaranteed it could become a solar power house in the near future. However, it also has an antipodal energy policy because it relies heavily on crude oil and related products for more than 36 percent of its exports. This trade generates desperately needed income for one of the poorest countries in the hemisphere, obligated to service a crushing debt-to-GDP ratio of up to 60 percent. In May 2023, Ecuador reduced its debt burden by $1.6 billion when it closed the world's largest debt-for-nature swap, in exchange for the financing of marine conservation activities in the Galápagos Islands area. To date, it has received $263 million from the Green Climate Fund but imagine how truly sustainable it could be if an adequate level of climate debt payments were to release it from the obligation to pump oil for export. If the Yasuní oil stays in the Amazonian soil, Ecuador will be the first country to "strand" its fossil assets and find a more sustainable pathway forward.

The "rights of nature" movement points in the direction of another way to strand assets. Its underlying doctrine seeks to shift our human relationship with nature from a property-based legal framework to an eco-jurisprudence that recognizes Earth's living systems as right-bearing entities. Aside from the constitutional victories in Ecuador and Bolivia (and almost in Chile), the most successful efforts to activate this philosophy have involved the protection of river and lake ecosystems, in Florida, Quebec, New Zealand, and India. The doctrine has also been invoked to ban fracking in a variety of US locations and has been championed by Indigenous communities in the ongoing struggle against pipelines. Typically, in the mining industries, rights are allocated to extractors. If ecosystems have an inherent and a priori right to exist and flourish, then these property rights can be overridden. The wild, but perfectly logical, thought then arises: do oil, gas, and coal deposits have a right to remain, unexploited, in the soil? Formed from ancient organic waste over tens of millions of years, they perform many functions in the balanced ecosystems of which they are a part and cannot be considered separable elements. Their extraction disrupts that ecological balance, while the process and products of refining add immeasurably to the destruction of natural environments. A legal system of Earth jurisprudence, designed to protect the health of planetary life, would rescue these fossil deposits from their foul reputation as climate culprits and restore their status as an inalienable part of nature.

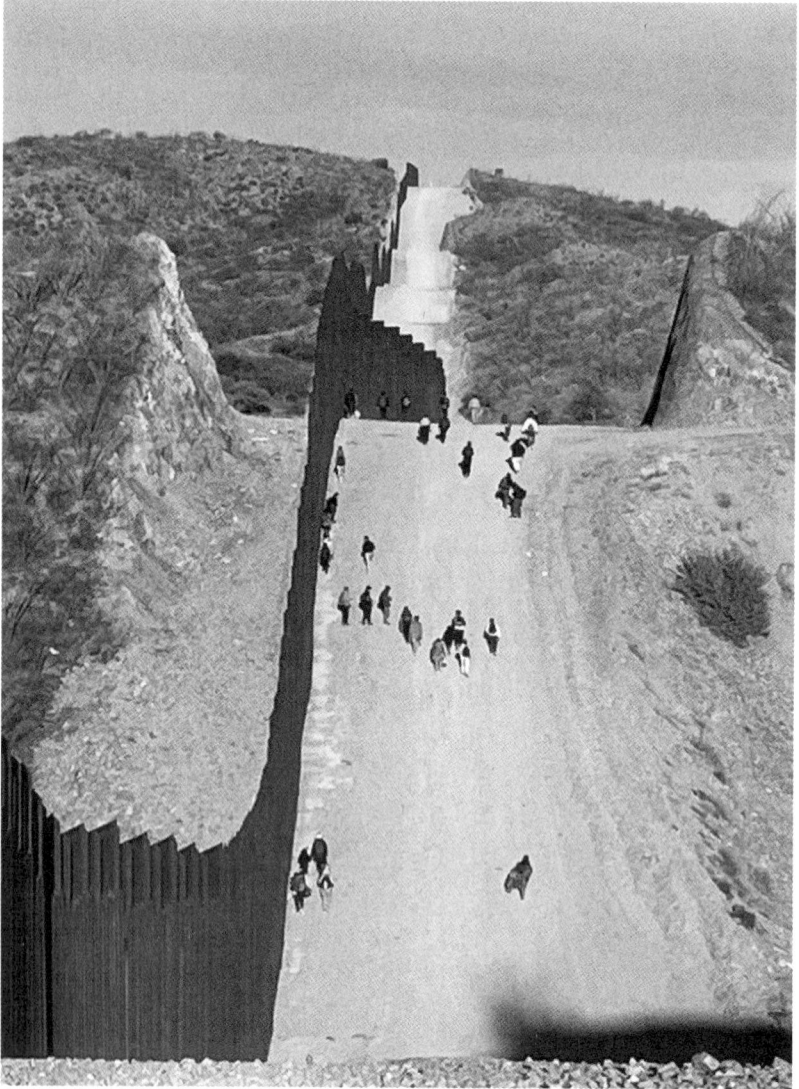

US-Mexico border wall, Sasabe, Arizona, February 2024.

3

ARIZONA

WHEN YOU HAVE MORE THAN YOU NEED, BUILD A LONGER TABLE, NOT A HIGHER WALL

A few months after I turned in my book, *Bird on Fire*, nominating Phoenix for the title of "world's least sustainable city," Jon Talton, the city's venerated crime writer, published a novel that featured a character loosely based on me. In the opening pages of *South Phoenix Rules*, Jax Delgado is introduced as an East Coast academic—with exactly my academic job description—who is researching sustainability in Arizona's desert metropolis. "That'll be a short paper," quipped the novel's protagonist, deputy David Mapstone. Shortly thereafter, Delgado's severed head is found in a box mailed to his girlfriend. It turns out that his scholarly identity was an alias. A high-ranking hitman for the Sinaloa narco-trafficking cartel, he was now a casualty in a war whose killing fields were migrating from Juarez to Phoenix.

The setting for Talton's novel, published in 2010, was the cross-border trade through which drugs moved north and high-grade weapons traveled in the other direction. The collateral damage generated by this two-way traffic was one of the keys to the fight over immigration, for which Arizona was a flashpoint at the time. Having passed the strictest immigration laws in the country, right-wing vigilantes, under the aegis of Maricopa County Sheriff Joe Arpaio, were busy earning the state a reputation for hatred of Brown people. Shortly thereafter, Donald Trump feasted on the popular animus against immigrants through his nonstop vows to build a "big, beautiful wall." But in an immensely gratifying turn of events, Maricopa County's Latinx organizers mobilized for years to take down Arpaio and then helped to turn the state blue in the 2020 election that deprived Trump of a second term.

In the intervening years, Texas stepped up to outdo Arizona in its zeal to punish border-crossers by busing migrants to Washington, DC and other "sanctuary" municipalities and by wresting its border from federal authority, denying access to agency officers and erecting its own concertina barbed wire. The transports were a way of declaring that "every state is now a border state," while the wire was grotesque political theater.[1] Other red states took the bait and explored a suite of measures to criminalize immigrants on the legal pretext that they constituted a literal "invasion" force.[2] In the run-up to the 2024 election, the MAGA faithful once again fixated on the border as the ticket to the White House. Trump was back, ranting in high Hitlerese about how immigrants were "poisoning the blood of our country,"[3] and his rallies were festooned with "Mass Deportation Now!" signs borne by his fans. Despite the predictions reflected in Talton's novel, and the MAGA alarmism about an invasion of criminal immigrants, the bloody mayhem of the cartels had not reached Phoenix, but the violence in Mexico, now awash with guns carried south from Wal-

---

1 Texas also hosted the most violent anti-immigrant attack, when a white supremacist killed twenty-three people in a Walmart Supercenter in the border town of El Paso in 2022. See Betsy Hartmann's analysis in *The America Syndrome: Apocalypse, War, and Our Call to Greatness* (New York: Seven Stories Press, 2023).
2 Jazmine Ulloa, "G.O.P.-Led States Claiming 'Invasion,' Push to Expand Power to Curb Immigration," *The New York Times*, June 15, 2024, https://www.nytimes.com.
3 Trip Gabriel, "Trump Escalates Anti-Immigrant Rhetoric with 'Poisoning the Blood' Comment," *The New York Times*, October 5, 2023, https://www.nytimes.com.

mart or Dick's Sporting Goods stores, was one of the factors driving people across the border in record numbers. Moreover, the cartels and corporations invested in the border industrial complex were drawing more profit from trafficking humans than from drugs. Either way, the politics of "securing the border" was guaranteed to be a priority in the presidential race, with the settler-colonial history of violent frontier conflict as its backdrop.

By 2023, Arizona was once again seeing the country's highest volume of border-crossers; up to 1,000 asylum-seekers a day were being processed by Department of Homeland Security (DHS) agents, while others tried to break through the formidable 100-mile security zone, with its multibillion dollar arrays of electronic surveillance, and trek across the unforgiving desert terrain without being apprehended.[4] Some of the migrants, hailing from as far away as China, India, Nepal, Russia, Eastern Europe, and West Africa, had traveled for thousands of miles, including through the treacherous Darien Gap jungle, or through a series of air jumps, from Istanbul to Ecuador, Colombia or Nicaragua, and had "tolls" exacted by cartels, and fees and shakedowns imposed by their smugglers. After such hardships, the Sonoran Desert and the Border Patrol checkpoints were just one last set of obstacles along the way.

In December 2023 alone, there were a record 300,000 "migrant encounters" in the four states with Mexican borders (9,600 per day) and many more uncounted crossings. As many as 3.3 million asylum cases were pending, with fewer than seven hundred immigration judges to adjudicate their claims. In an effort to head off a looming electoral liability, the Biden administration, which had failed miserably to implement a humane immigration policy, leaned into measures that beefed up entry restrictions, eventually "closing" the border to asylum seekers in June. On a different front, looking to stem his hemorrhaging support from climate advocates, another key voting constituency, the White House placed a pause on approvals for new fossil fuel export facilities.

---

4   Oscar Martínez, *The Beast: Riding the Rails and Dodging Narcos on the Migrant Trail* (New York: Verso, 2014).

These two issues were twinned in the MAGA mind; the most favored policy chants at Trump rallies were "Deport Them All" and "Drill, Baby, Drill." But federal policy did not have much to say about connecting them. Indeed, none of the conditions under which refugees could qualify for asylum (including modifications to the UN's Refugee Convention of 1951) took climate change into account. And yet the ties between global warming and immigration have been crystal clear for a long time. In *Bird on Fire: Lessons from the World's Least Sustainable City*, I pointed out that a significant fraction of Arizona's border crossers could fairly be classed as climate refugees, forcibly displaced by warming impacts that had been disproportionately brought on by the carbon emissions of Arizonans' high-carbon lifestyle. Surely, they had a right to safe harbor along with reparative benefits for their losses? There wasn't much of a response to this argument at the time.

In the intervening years, the climate impacts had intensified. Campesino livelihoods were devastated by a long drought in Central America's "dry corridor," encompassing parts of El Salvador, Guatemala, Honduras and Nicaragua, countries that are already among the world's most vulnerable to climate-driven damage.[5] Refugees from these parched areas were entering a US border state suffering through a mega-drought that began in 1998 and showed no signs of letting up. They were also coming to a country that had seen its own rate of internal climate migration soar; as many as 3.2 million have had to abandon their homes as a result of flooding alone.[6] Worldwide, as many as twenty-one million climate migrants have been displaced on an annual basis since 2008, according to the United Nations High Commissioner for Refugees.[7] In 2024, the International Organization

5   Jonathan Blitzer, "How Climate Change Is Fueling the US Border Crisis," *New Yorker*, April 3, 2019, https://www.newyorker.com.

6   Aimee Picchi, "About 3 Million Americans are Already 'Climate Migrants,'" *CBS News*, December 18, 2023, https://www.cbsnews.com/; Abrahm Lustgarten, *On the Move: The Overheating Earth and the Uprooting of America* (New York: Farrar, Straus and Giroux, 2024).

7   United Nations High Commissioner for Refugees, https://www.unhcr.org/uk/news/stories/frequently-asked-questions-climate-change-and-disaster-displacement.

for Migration reported that climate change was a primary cause for the "roughly 281 million people worldwide," or 3.6 percent of the global population, who are on the move.[8]

Drought, which afflicts one-third of humanity, is one of the biggest drivers of displacement around the world. In the dry corridor states, and also in Mexico to the north, irregular precipitation patterns, rising temperatures, corporate monocropping, and increased storm intensity have also taken a savage toll on small-scale arable production. So, too, it is far from easy to clearly distinguish climate impacts from other causal factors—such as poverty, insecurity, and intimidation—that set migrants in motion northwards, although there is no doubt that it is amplifying their inability to stay in their homes. Many rural and Indigenous families were displaced due to free trade agreements and decades of US-promoted development policies that favor large-scale agricultural producers of cash crops for export. Others had fled the US dirty wars in Central America and were now seeking refuge from the cartel violence that had spread to many Mexican states.[9] But no one can dispute global warming's footprint on the hemisphere, and it is getting larger every year. As Erick Meza, the Borderlands coordinator for Arizona's Sierra Club chapter, put it, "weather patterns are changing faster than people can adapt with the resources they have." Meza sees the border, and the people crossing it for survival, as a "thermometer that measures the temperature of

---

8   International Organization for Migration, *World Migration Report 2024*, https://worldmigrationreport.iom.int/msite/wmr-2024-interactive/. A World Bank report has calculated that by 2050 there will be at least 216 million internal climate refugees, moving within their own countries. Viviane Clement et al., "Groundswell Part 2: Acting on Internal Climate Migration," World Bank (Washington, DC, 2021), http://hdl.handle.net/10986/36248.

9   Daniel Faber, *Environment Under Fire: Imperialism and the Ecological Crisis in Central America* (New York: Monthly Review Press, 1993); Aviva Chomsky, *Central America's Forgotten History: Revolution, Violence, and the Roots of Migration* (Boston: Beacon Press, 2022); Jeremy Slack, *Deported to Death: How Drug Violence is Changing Migration on the US-Mexico Border* (Berkeley: University of California Press, 2019); Jeremy Slack et al. (eds.), *The Shadow of the Wall: Violence and Migration on the US-Mexico Border* (Tucson: University of Arizona Press, 2018); and Jonathan Blitzer, *Everyone Who Is Gone Is Here: The United States, Central America, and the Making of a Crisis* (New York: Penguin, 2024).

what's happening in their countries of origin, and if things are really bad there, you will see it reflected here," adding, "how bad do these things really have to be to make you risk this journey?" Another Tucson resident, Todd Miller, a well-known and prolific "border writer" whose analyses of climate impacts on migration pull no punches, has memorably described the technologically enhanced border wall as "a climate adaptation program" for those who are lucky enough to be born on the northern side of it.[10]

Despite a rocky history, environmental activists and immigrant advocates have found common cause over the predicament of climate refugees, with both camps pushing the rights claims of environmental migrants into the domain of international legal protection, albeit with limited success so far.[11] But white supremacists have also seized the opportunity to upgrade their racial dog whistles by co-opting environmental messaging. From the early 2000s, border vigilante groups were making noises about the damage done to fragile desert ecosystems by migrants, while turning a blind eye to the devastation caused by Customs and Border Patrol (CBP) movements on wildcat roads and by the massive construction of DHS infrastructure within the border security zone. In 2021, Arizona's right-wing Attorney General Mark Brnovich amplified these complaints by suing the DHS on the basis that the Biden administration's immigration policies had "directly result[ed] in the release of pollutants, carbon dioxide, and other greenhouse gases into the atmosphere, which directly affects air quality."[12] Such language, suggesting that immigrants caused pol-

---

10 Todd Miller, *Storming the Wall: Climate Change, Migration, and Homeland Security* (San Francisco: City Lights, 2017), 42.

11 In 2018, leaders from 164 countries formally adopted the UN Global Compact for Migration at a meeting in Marrakesh, Morocco. The agreement identified climate change as a driver of migration and urged countries to start planning to accommodate the movement of climate refugees. But the pressure to win legal refugee protection has encountered many obstacles. See Jane McAdam's summary, "'Protection' or 'Migration'? The 'Climate Refugee' Treaty Debate," in *Climate Change, Forced Migration, and International Law* (Oxford: Oxford University Press, 2012).

12 Alexander Kaufman, "How Arizona's Attorney General is Weaponizing Climate Fears to Keep Out Immigrants," *Grist*, May 4, 2021, https://grist.org.

lution, was a barely veiled way of saying that immigrants themselves were pollutants.

Virulent rhetoric about the biological threat of immigration has a long history in the US, dating from the eras of racial science and eugenics, but it took on a new lease of life with the birth of the modern environmental movement. Panic about the ecologically defined threat of overpopulation almost inevitably centered on the fecundity of Brown people from the South. A succession of biologists, ethnologists and demographers—William Vogt, Paul and Anne Ehrlich, Garret Hardin, Kingsley Davis, John Calhoun, Desmond Morris, and David Attenborough—stoked up neo-Malthusian fears that the planet's carrying capacity would collapse unless drastic measures were taken to reduce populations, both in the developing world and among minority groups in the North with larger families. These fears were parlayed by ophthalmologist John Tanton into an energetic network of anti-immigrant groups—Federation for American Immigration Reform, Center for Immigration Studies, Zero Population Growth, NumbersUSA—with a laser-focus on passing legislation about border exclusion.[13] Most recently, the "greening" of these nativist organizations has given rise to eco-fascist politics, especially among resurgent far-right European parties in Germany, France, Italy, Poland, and Hungary.[14] "Blood and soil" ideology has returned to the front burner,

---

13 Reece Jones places this movement within a larger history of racial exclusion at the border in *White Borders: The History of Race and Immigration in the United States from Chinese Exclusion to the Border Wall* (Boston: Beacon Press, 2021).

14 Among the European *ur*-texts are Jean Raspail's dystopian novel, *Le Camp des Saints* [*The Camp of the Saints*] (Paris: Editions Robert Laffont, 1973) and Renaud Camus, *Le Grand Remplacement* [*The Great Replacement*] (Chez L'Auteur, 2011). An English-language influence is Alan Brimelow, *Alien Nation: Common Sense About America's Immigration Disaster* (New York: Random House, 1995). For critical analysis, see Janet Biehl and Peter Staudenmaier, *Ecofascism: Lessons from the German Experience* (San Francisco: AK Press, 1995); Sam Moore and Alex Roberts, *The Rise of Ecofascism: Climate Change and the Far Right* (London: Polity, 2022); John Hultgren, *Borders Walls Gone Green: Nature and Anti-Immigration in America* (Minneapolis: University of Minnesota Press, 2015); Blair Taylor, "Alt-Right Ecology: Ecofascism and Far-Right Environmentalism in the United States," in *The Far Right and the Environment*, ed. Bernhard Forchtner (New York: Routledge, 2019).

with white nationalists professing to holding the line in their defense of natural patrimonial heritage, ethnic purity, and "home-grown" culture. Adding environmental contamination to the "great replacement" theory reinforced the belief that the homeland and its rightful, rooted custodians are under threat of inundation, not by rising seas but by foreigners. In this fevered scenario, the homeland is imperiled, not by climate change but by outsiders pouring across borders, bent on "reverse colonization."[15]

By the time of Trump's third presidential run, the foul stew of allegations about an invasion force of over-breeding, contagious migrants from what he had called "shithole countries" formed a choral accompaniment to his own ethnonationalist obsession with the border. Appropriately, in his first administration, he had used an eighty-year-old public health law to close off the border during the COVID pandemic. That did not stop migrants from crossing any more than his earlier "zero tolerance" policy on family separation did. Since then, the budgets for the involved agencies—ICE and CBP—had all skyrocketed, to $30 billion by 2024, with little effect; no massing of officers or upgrades in security technology had kept people from coming. As for the built sections of his "big, beautiful wall," the adage on the ground was that "it stopped everything, except humans."

What Betsy Hartmann once diagnosed as the "greening of hate" has evolved over time into an arsenal of white supremacist talking points about how and why the granting of asylum will upset the balance of nation and nature.[16] In some respects, we are not far removed from Malthus' crusade against Britain's Poor Laws at the turn of the nineteenth century. Malthus argued that any further governmental efforts at direct poor relief would interfere with the nation's "natural" checks on population growth—primarily the Christian self-restraint of its middle class. A surge in numbers would strain resources, he reasoned, and encroach on the food supply reserved for those wealthier and more deserving citizens from superior racial stock. More than two centuries later, at a time when overt racialist promotion is no longer acceptable, the same arguments are being cloaked in a pseudo-

---

15 See Malm and the Zetkin Collective, *White Skin, Black Fuel*.
16 Hartmann, "The Greening of Hate: An Environmentalist's Essay."

environmentalist mantle. If rich nations absorb population surpluses from the Global South, we are warned, global warming will be magnified when their oversized families adopt high-carbon lifestyles. In an era of climate change, the nation's stock of natural resources is already under strain, and it will be further imperiled by foreigners from different ecosystems.

None of these green nationalist talking points about immigration has any basis in fact. Their appeals to environmental science are as gaseous today as the ones to racial science in an earlier century. But it does not take much to tap into popular anxiety about scarce resources—the always available dread that there simply isn't enough to go around. It is this kind of fear that gives rise to a hoarding mentality, where assets are plundered from the commonwealth and squirreled away behind gates, walls, and LLCs. The stockpiling is clearly reflected in patterns of eco-apartheid in cities, like Phoenix, where the well-heeled enjoy cleaner air, purer water, and access to quality nutrition and green gizmos, while those on the other side of the tracks live in food deserts, on floodplains, with no tree cover and in proximity to polluting industrial plants.

Arizona's citizenry was prone to this kind of unease. For one thing, it is a border state, and it hosts the most crossings in the nation. More importantly, concerns about scarcity, of water above all, are ingrained. All of the patterns of Southwestern settlement, and the lifestyle choices of settlers, have been shaped by the limited water supply. The future growth of the region hangs, perennially, on the need to find "the next bucket" of water. As a result, there have always been doubts about the capacity to sustain a large population in the country's hottest desert, with an average annual regional rainfall of only eight inches. Overpopulated is a stigma that Central Arizonans have borne for a long time, though it has not stopped citizens ("amenity migrants") from coming from other parts of the US. From year to year, Maricopa County records more arrivals than any other county in the nation, attracted primarily by housing prices that are still lower, at least on the urban fringe, than any other metro area.

It is no surprise then that a migrant surge from the South would trigger a population scare in a state already anxious about its ability to absorb newcomers. Despite the evidence that most migrants were bound for destinations far from Arizona, fears about their presence, inflamed by race hatred and xenophobia, trouble those who wonder whether the region can support larger numbers of people. The question isn't simply one that resonates with the "incumbents' club." It has to worry anyone who believes there will not be enough water to sustain the existing population, let alone the thousands of newcomers from other parts of the US for whom freshly built tract homes lie in wait just beyond the urban fringe. These homes are the latest—and undoubtedly the last—crop yielded by lands that have managed, against the odds, to host desert farming for at least a millennium and a half.

LEAKY BUCKETS

It won't be recorded as an official milestone in the state's history, but the governor's decision to deny construction permits for new subdivisions on the fringes of the Phoenix metro region in June 2023 was as earthshaking as anything the state had seen. The shockwave set off alarms in the C-suite offices of the nation's merchant builders, and it was felt in other drought-stricken Southwestern states where settlement and housing growth had banked on finite groundwater supplies. The announcement followed on the release by the new Democratic governor, Katie Hobbs, of an Arizona Department of Water Resources (ADWR) report that had been suppressed by her Republican predecessor. According to the agency, long-term estimates of physically available groundwater in outlying suburban areas showed a significant amount of "unmet demand" (or future shortage) of 4.86 million acre-feet—which means that the remaining aquifer reserves had been fully allocated and could not supply any more need.[17] Under the state's Assured Water Supply regulations, approval certificates could only be issued to developers if they could show there would be adequate supplies for 100 years. It was by far the strictest requirement in the country, but decades of overdraft had depleted the region's sub-basins and threatened aquifer stability. Now there were insufficient guarantees of water in the years to come to justify

17 Phoenix AMA Groundwater Supply Updates, Arizona Department of Water Resources (2023), https://www.azwater.gov/phoenix-ama-groundwater-supply-updates.

digging new subterranean wells for housing projects in outer suburbs of Phoenix's East and West Valley areas.

The governor's order to deny the permits was long overdue, and it was legally required in order to ensure the state's compliance with the far-seeing Groundwater Management Act (GMA) of 1980, which sought to limit the use of nonrenewable groundwater. But the growth machine that had churned out sunbaked housing tracts for so long and had only just recovered from the pandemic pause would not be grinding to a halt any time soon. Urban areas already serviced by municipal water companies with assured supplies were not affected by the governor's order, existing water certificates in fringe areas would be honored, and developers could still build if they found "alternative water sources." That meant they would have to find some way of tapping into the state's allocation of Colorado River water, which is pumped, at great cost, three hundred miles uphill through the Colorado Arizona Project (CAP) canal to service the cities of Phoenix and Tucson and the rich farmlands in between. Developers could purchase CAP rights from the region's farmers or from the several Native American tribes who have the most longstanding water rights. Those alternative sources would not come cheap, however, and would inevitably stymie the low-cost housing formula that reliably draws newcomers to the region.

Drawing on decades of proven success in buying off elected officials, the all-powerful Home Builders Association of Central Arizona sprang into action to challenge what it provocatively called the governor's "moratorium on home building in the most affordable parts" of the region. The same restrictions were not being applied to growers, the association pointed out, and besides, developers who built on "retired" farmland would be using much less water than the quantities that thirsty crops like alfalfa and cotton required. The association's complaint that its members were being unfairly targeted cited other loopholes.[18] The water supply requirements did not apply in rural lands outside the "active management areas" set up under the GMA, and the burgeoning build-to-rent sector of the real estate industry

---

18 Howard Fischer et al., "Homebuilders Want Changes to Ease Water Law Restrictions," *Arizona Capitol Times*, December 21, 2023, https://azcapitoltimes.com.

was also exempt. These were warning shots across the bows of the Democrats' fragile ship of state, helmed by the governor, attorney general, and secretary of state.

In the meantime, the legislature's hard-right Republican majorities could be counted on to deliver more firepower in the months ahead. They were not about to let Arizona's premier economic engine—once described to me as "people building homes for people who are building homes"—be sacrificed at the liberal altar of sustainability. Central Arizona's awesome growth machine had faced such obstacles before, but this time around, its beneficiaries would be bumping up against hard facts about hydrological scarcity that could be more difficult to circumvent. Reflecting the GOP's nationwide lurch to the far-right, Arizona's fringe Republican lawmakers now occupied the mainstream, and were busy, as usual, introducing batshit crazy bills, few of which would move out of committee. But they were deadly serious about protecting the real estate industry, and so, as the 2024 session got underway, there were almost a hundred water bills vying for passage in both chambers.

In 2009, when I finished *Bird on Fire*, construction was at a standstill, due to the financial crash brought on by speculation on mortgage loans. Tumbleweed rolled down the empty streets of Arizona's newest ghost towns, about to be reclaimed by the desert on the urban fringe. After 2014, banks reopened their lending arms, business as usual resumed, and the industry leaned into its profitable formula of spitting out subdivisions of single-family homes. The 2020 census showed that Phoenix, the country's fastest-growing city, had added a quarter of million homes in the last decade, while its twenty-one satellite cities kept pace. New water conservation regulations—the strictest in the country—had resulted in an overall decrease in per capita use, but every drop saved was then claimed for the new developments. This was the paradox that lay at the heart of the West's transactional water politics. Despite the conservation measures, there were more and more skirmishes over water regulation, and anxiety about shortages began to build. Five months before the governor's announcement, the suburban city of Scottsdale turned off the water pipe to the nearby community of Rio Verde Foothills, leaving most residents with little option but to pay for costly deliveries from water tankers. An interim solution was eventually cobbled together, but the spectacle of a desirable desert town running dry was widely seen as a symptom of times to come when triage management could determine which communities will get cut off.

This lifeboat calculus had already taken a large bite out of the state's assigned allocation of CAP water. Under the Colorado River Compact, which determines the shares of seven states and 40 million people, Arizona held "junior rights." This disadvantage made it first in line for cuts, and so, in 2021, when the federal government declared a Tier 1 shortage, the state's share of the depleted river's water was reduced by 21 percent, amounting to 9 percent of the state's overall use. The strain on resources was determined from the outset by the compact's original overestimate of the river's flow in 1922. But, more recently, years of low snowpack had left the telltale bathtub ring in Lake Mead increasingly exposed, as the reservoir that supplied the Lower Colorado Basin states came close to a critical "dead pool" condition when downstream flow would cease. To prevent a further mandatory Tier 2a reduction in 2023, the lower states submitted to additional voluntary cuts, but could not agree among themselves to the ratios, so the feds stepped in to sweeten the pot by using $1.2 billion of Inflation Reduction Act (IRA) funds to compensate stakeholders, especially farmers who are the first to get cut off because they have the weakest rights. Like it or not, the federal government owns the entire water system in the Western US and is the only entity that can broker a deal between warring states.

But another player of significance had emerged of late. Even though they hold "senior rights" that are federally reserved through treaties and are entitled to more than 45 percent of the state's CAP water, Arizona's twenty-two registered Indigenous tribes had not been regarded as equal partners in negotiations until recently.[19] Tribal nations were barely mentioned in the original 1922 Colorado River Compact. As part of the 2023 voluntary settlements, the Gila River Indian Community (GRIC) and the Colorado River Indian Tribes contributed a large share of their own allocations in return for fed-

---

19 Debra Utacia Krol, "Tribes Play a Greater Role in Water Management as Drought and Climate Change Effects Worsen," *Arizona Republic*, August 8, 2022, https://www.azcentral.com.

eral funds. The new restrictions thrust them into pole position as *protectors* of Lake Mead, and, by extension, of Phoenix and Tucson. Tribal leaders willingly accepted the gratitude, along with the newfound respect. But they made it clear that they would not be at the table solely to negotiate over water as a transactional commodity to be bargained on behalf of short-term economic goals; they would also be approaching the conversation about drought management from the "spiritual and cultural" perspective of Indigenous values.[20] "Water is Life" means something different to the stewards of Turtle Island than to project managers trying to secure certification for their company's next thirty cookie-cutter single family homes.

At the same time, tribal leaders were using their newfound central role to convert their "paper rights" into real water to be delivered through an adequate network of pipes and canals. Having waited so long for legal recognition of their access, they were seizing the opportunity to extract the necessary infrastructure funding. GRIC fought for more than eighty years in the courts before winning the right, through a momentous 2004 settlement, to 600,000 acre-feet per year, 311,000 of which is CAP water (more than the city of Phoenix). With this entitlement, they were aiming to fully restore their former prosperity as irrigation farmers of the region's "breadbasket" before upstream settlers diverted the powerful Gila River flow and consigned the tribe to destitution in the 1910s. Indian communities closer to the Colorado River mainstem were restoring riparian zones and estuary lands with the goal of building resilience for the river and for themselves. There may come a day when the 1,450-mile-long river—which has not reached its Mexican delta since 1984—will once again flow into the sea, but it will probably take the demolition of some dams to make that happen.

For the best part of the twentieth century, the Bureau of Reclamation and its sometime rival, the Army Corps of Engineers, had been rearranging the plumbing of almost every river in the West to ensure that Anglo settlers could subsist on their Homestead Act farms or in cities, like Tucson, that relied entirely on pumped groundwater

---

20 Debra Utacia Krol, "Their Pleas for Water Were Long Ignored. Now Tribes Are Gaining a Voice on the Colorado River," *Arizona Republic*, August 8, 2022, https://www.azcentral.com.

until 1992. Among the outcomes were engineering feats, mostly in the form of mighty dams, that established American technological prowess in the world at large. Overseas, the dams were the signature mark of Western capitalist modernity, as developing countries received generous, albeit conditional, funding from the World Bank and IMF to build the hydrological monsters. The benefits included cheap hydropower and flood control, but almost everywhere, the projects were plagued by hyper-salinization, silting up, and riparian degradation, with small farmers, Indigenous communities, and riverine wildlife paying the highest price.

By the end of the century, Indigenous resistance won a verdict from the World Commission on Dams that the environmental and social costs had been too high. The un-damming movement kicked into gear. Of the 75,000 dams in the US, more than 2,000 have either been removed or slated for demolition to restore salmon migration and damaged ecosystems.[21] In February 2024, environmentalists and tribal communities celebrated a milestone—the dynamiting of the last of four decommissioned dams on the powerful Klamath River in Oregon and Northern California, making way for the revival of its riparian corridor.[22] But so far, no one was talking about demolishing the more indispensable dams of the "dry Sunbelt," especially those on the Colorado or the Salt-Verde system that watered Central Arizona. Tens of millions of settlers depended directly on the careful hydraulic management of their flow.

The Bureau's long season of Big Water projects may have been the last publicly funded wave of settler colonialism, following the American Indian Wars and the legislative chicanery that cheated tribes out of 2 billion acres of their land. Megatons of concrete were poured to shore up the geography of Anglo domination and give material form to the power hierarchy of the West.[23] The Reclamation

---

21 See Elizabeth Grossman, *Watershed: The Undamming of America* (New York: Counterpoint, 2002).

22 Debra Utacia Krol, "As Klamath River Dams Come Down, Old Lands Emerge and the Landscape Begins to Heal," *Arizona Republic*, February 16, 2024, https://www.azcentral.com.

23 F. Lee Brown and Helen Ingram (eds.), *Water and Poverty in the Southwest* (Tucson: University of Arizona Press, 1987).

Act of 1901 that instituted the Bureau allowed unceded Indian land
to be seized by eminent domain for these vast infrastructure projects.
According to one study, more than 1.13 million acres of tribal land
were flooded under the reservoirs of 424 dams alone, amounting "to
an area larger than Great Smoky Mountains National Park, Grand
Teton National Park, and Rocky Mountain National Park combined."[24]

Indian nations were given senior water rights as part of Western settlement treaties in the nineteenth century, but they were not
worth much at the time. All that has changed now that the Colorado's
quarter-million-square-mile watershed is delivering less and less to
the river. Once they reach a legal settlement, the twenty-nine federally recognized tribes in the Colorado River Basin will hold rights to
a combined 3.6 million acre-feet—roughly 25 percent of the river's
annual flow.[25] These entitlements are arguably more valuable today
than the minerals found underneath lands allocated for Indian reservations because they were considered agriculturally unproductive.
Extraction rights for these minerals were traded away to mining companies from a position of weakness, even after tribal councils established firmer self-control. Water rights are proving to be a different
story. Given the colonial underpinnings of Western hydrology, it is a
remarkable exercise of environmental justice that Indian tribes may
end up as key water providers—one of the most powerful roles in the
West. This has happened at a time of Indigenous re-assertion; after
the 2016 Standing Rock protests, "Water is Life" became a rallying cry
for the defense of Indigenous cultural practices and spiritual values.

ARIDIFICATION IS THE NEW NORMAL

As Marc Reisner showed in *Cadillac Desert*, his classic account of
Western irrigationism, the dams that made the Bureau of Reclamation's name were the greatest, and possibly the most permanent,

---

24 Heather Randell and Andrew Curley, "Dams and Tribal Land Loss in the
United States," *Environmental Research Letters* 18, no. 9 (August 2023).
25 Leslie Sanchez et al., "The Economics of Indigenous Water Claim Settlements
in the American West," *Environmental Research Letters* 15, no. 9 (August 2020).

public works projects ever tried in the US.[26] In this regard, they were quasi-socialist exercises in the nationalization of water supply, under-taken after small-scale capitalist efforts failed, and they turned the West into a dependency of federal authority.[27] The region's strident libertarians love to inveigh against the state but their cities would not last long if government agencies withdrew their responsibilities for managing the distribution of water flows. The Anglo founders of Phoenix built an oasis settlement below the confluence of the capricious Salt and Verde rivers, but its mid-century growth spurt was only assured by the Bureau's upstream construction of a series of dams for flood control in the preceding decades.

Writers with a taste for "collapse porn" are fond of imagining what would happen if federal oversight crumbled. In his near-future novel *The Water Knife*, Paolo Bacigalupi describes hardscrabble life in drought-ravaged Phoenix after CAP has been blown up by saboteurs, suspected to be agents of a Los Vegas cartel. California and Nevada are in a low-intensity conflict with Arizona over every last drop from the Colorado River. Government authority has ceded to organized crime, water rights are openly traded or stolen, and the corporate elite live in heavily guarded ecotopian towers, while residents who cannot afford to leave have to pay at the pump for buckets of the remaining groundwater.

But you do not need to be a fiction writer to project a grisly future for the region. In his 2011 book, *A Great Aridness*, conserva-tionist William deBuys drew on scientific research to suggest that the Southwest is in transition to a new climate norm of *aridification*, and that the current drought will be more or less permanent, not an exceptional, multi-year phenomenon with an expiration date. The impact of atmospheric CO_2 build-up means that there is no return to the stability of the Holocene era, notwithstanding its occasional bouts of protracted drought. As deBuys concludes, the "hydraulic society" so strenuously built over the last century and a half will have

---

26 Marc Reisner, *Cadillac Desert: The American West and its Disappearing Water* (New York: Viking, 1986).
27 See Donald Worster, *Rivers of Empire: Water, Aridity, and the Growth of the American West* (New York: Pantheon Books, 1985).

to "reinvent itself in new and undoubtedly painful ways."[28] Along the way, like so many other commentators on the "dry Sunbelt," he ponders the fate of prehistoric Southwestern cultures like the Ancestral Pueblo (Anasazi), Mogollon, and Hohokam, whose cities were abandoned in the face of prolonged drought, and the warfare occasioned by it. Before them, in the world's prehistory, lay the ruins of countless other ancient, hydraulic civilizations, whose demise was spurred by drawn-out changes in climate, especially those settlements which flourished in deserts, where the desolation is always more calamitous.

The most locally illustrative were the Hohokam (or "the departed ones"), who farmed the Phoenix Basin for more than a millennium. They controlled and diverted the waters of Salt and Gila Rivers through an extensive system of distribution canals, lateral networks and ditches, often with sophisticated sluices and gates. At peak population, the Hohokam irrigated more than 110,000 acres and supported small cities and far-flung trade until the culture became archaeologically invisible in the mid-fifteenth century. No other US city is built on top of a collapsed pre-Columbian civilization, and the Hohokam downfall has long served as a morality tale for Phoenicians living high on the hog. Yet many of their descendants—including the Akimel O'odham and Posh Pee on the Gila River Indian Reservation (GRIC)—are now key to the region's future. GRIC's recently won water rights (entitling them to a quarter of all of the state's CAP water—a larger share than the entire state of Nevada) are not subject to the restrictions placed on other Arizona residents, and their community's respect for the cultural sanctity of water sets them apart from the other lawyered-up transactionalists who compete with each other in the Western water market.[29]

Despite the recent ascendancy of the tribal nations, there is little desire to play the sacrificial savior for settlers whose faith in the long-term availability of water deBuys likened to believing in the Tooth Fairy.[30] Indigenous people have long been regarded as "solution-

---

28 William deBuys, *A Great Aridness: Climate Change and the Future of the American Southwest* (New York: Oxford University Press, 2011), 9.
29 Rachel Monroe, "How Native Americans Will Shape the Future of Water in the West," *New Yorker*, January 27, 2023, https://www.newyorker.com.
30 DeBuys, *A Great Aridness*, 172.

providers" for Anglo settlers, especially environmentalists who tend to hail Native Americans as ecological stewards. Yet these exchanges have mostly been one-way; there have been few solutions on offer for Indian problems. Because of their superior rights and their integrity as "water defenders," tribal nations like GRIC, whose lands lie directly in the path of Phoenix's expansion to the south, occupy the moral high ground. GRIC leaders showed they were responsible stakeholders by electing to leave much of their allocated water (by 2024, as much as 50,000 acre-feet) in Lake Mead, but they also insisted that everyone cooperated in taking a hit, when they helped to broker the drought contingency plan in 2019. Above all, they are now in a position, after decades of being cut off from the Gila River, to ensure their own self-determination.

As part of its 2004 settlement, GRIC was required to lease some of its water to nearby Mesa and Gilbert, but it could also offer more leases on a voluntary basis. Over the years, the community has entered into agreements with other cities like Chandler, Phoenix, Goodyear, and Apache Junction, utilities like the Salt River Project, and even mining companies. Some GRIC members still bristle at the idea of selling water credits because it appears to run counter to tribal values. As one young community member said to me, "We waited for eighty years to get our water back, and now we sell it back to people who stole it from us? That's not what Indians are about." David DeJong, the erudite director of the Pima-Maricopa Irrigation Project (P-MIP), told me that they are always looking for "a balance between the economic benefits to the community and its cultural values." P-MIP is a Bureau of Reclamation-funded agency charged with restoring and expanding GRIC's canal distribution system. Initially, thirsty municipal neighbors thought that GRIC would not use much of their CAP allocation, but P-MIP decided to take more water than they needed and store it underground. By recharging the aquifer in this way, GRIC claims more of its due share and banks it for future recovery. After all, "use it or lose it" is the mantra of Western water. After a year's storage, GRIC also earns credits that can be traded to others.

At one of the recharging sites, essentially a bank account, or "cash register," for the community, DeJong showed me P-MIP's effort to divert some of the water to recreate a permanent portion of the Gila River, which usually only runs for between one and three weeks

a year. This year it had not run at all, but the runoff from P-MIPs
riparian restoration has brought back cattails, cottonwood, willow,
and other species used by GRIC's artisans for weaving and traditional
medicines. This partial recreation of the river was a feat of great cultural significance to the community, and DeJong took visible pride in
the role his agency was playing "in righting the historical injustice"
that had taken the river away, more than a century before.[31]

A more sensitive topic is the resistance of some of GRIC's farmers to changing their crop mix. Alfalfa, the top cash crop, is very profitable. Easy to grow and tend, it can be harvested eight times a year
for feedlot use. But this thirsty crop contributes nothing to GRIC's
food self-sufficiency and it is labor-intensive to grow the kind of produce that would do so. GRIC's aspiration to restore its status as the
region's breadbasket will take time and it will require more attractive
incentives for growers to diversify. Even so, the community's goal of
restoring 75,000 acres of farmland is on track—and young members
are aspiring to be farmers in ways unthinkable a generation ago. A
leader in Indian country on account of its water management, GRIC
is also one of the few places in the region where cultivated acreage is
actually on the increase. With the aim of providing renewable energy
to the whole reservation, P-MIP has also begun to install solar panels
along twenty miles of its canals. It is the first such initiative in the US.
Because land is considered sacred and off-limits, GRIC has rejected
offers from solar farm companies, but the canals are fair game. Some
of them are thousand-year-old Hohokam ditches and now they are
hosting twenty-first-century clean technology.

By selling credits from the stored water in its own aquifer,
GRIC's leaders were able to tell themselves that they were not phys-

---

31 David DeJong is the author of three scholarly books that recount how the
Akimel O'odham and the Pee Posh, formerly known as the Pima Indians, were
deprived of their water: *Stealing the Gila: The Pima Agricultural Economy and Water
Deprivation, 1848–1921* (Tucson: University of Arizona Press, 2009); *Diverting the
Gila: The Pima Indians and the Florence-Casa Grande Project, 1916–1928* (Tucson:
University of Arizona Press, 2021); *Damming the Gila: The Gila River Indian
Community and the San Carlos Irrigation Project, 1900–1942* (Tucson: University of
Arizona Press, 2024).

ically selling the water. The buyers would be extracting the water from another aquifer. "It makes no hydrologic sense," conceded De Young, "but the law says you can do this." Still, it took a while for tribal leadership to accept the arrangement; the more acceptable language is along the lines of "sharing with the neighbors." GRIC's water banking program, overseen by the Arizona Water Banking Authority, had been one of the ADWR's innovations to meet uneven needs and smooth out variability across the region. Any excess CAP water could be stored in the aquifer in the event of future shortages. Other innovations were devised to provide what historian Douglas Kupel describes as "fuel for growth."[32] The first was the GMA itself. Sandy Bahr, the longstanding director of the state's Sierra Club chapter, described the 1980 act as a "leaky bucket," since it left open so many loopholes, but she acknowledged that it was still an "important effort to live within our means." Unlike the states of Texas, Oklahoma, and Colorado, which allow groundwater to be pumped until wells run dry, Arizona's law aimed to protect its aquifers, or at least to encourage a "safe yield" that would not deplete them.[33]

After the GMA's groundwater restrictions were introduced, the real estate industry howled and then pushed back hard. In 1993, in a lavish gesture of appeasement, the state created the Central Arizona Groundwater Replenishment District (CAGRD), which is required to replace pumped groundwater by finding renewable water supplies and injecting that water back into the aquifer. CAGRD, the first of several "workarounds" that "weakened" the GMA in Bahr's view, was a gift to developers who could not satisfy the GMA's Assured Water Supply program's requirement to demonstrate availability of a hundred-year supply of water. Now they could pump on the site of their subdivisions, and simply pay a fee to CAGRD to replenish, at some later date, the same amount of water somewhere else. CAGRD was not even required to replace the water into the same aquifer.

---

32 Douglas E. Kupel, *Fuel for Growth: Water and Arizona's Urban Environment* (Tucson: University of Arizona Press, 2003).

33 Kathleen Ferris and Sara Porter, "The Myth of Safe-Yield: Pursuing the Goal of Safe-Yield Isn't Saving Our Groundwater," Kyl Center for Water Policy, Morrison Institute (Phoenix: Arizona State University Press, 2021).

From the outset, the scheme was regarded as a "shell game" designed to give cover to developers. Nevertheless, with the backing of the region's professional water managers and hydrologists, the logic of what ASU's Kyl Center for Water Policy called an "elusive concept" was enough to breathe new life into the growth machine.[34]

In the go-go years that followed, subdivisions sprouted in every direction, magnifying CAGRD's obligations. CAGRD made it possible to build on cheaper desert lands with no past history of water use, and so there was no net deficit from retiring farmland. As a result, the pressure on groundwater dramatically increased. So, too, the mega-drought conditions emerged not long after CAGRD was initiated, and enrollee fees have shot up as surface water becomes increasingly scarce. Facing an over-enrollment crisis in a time of deepening drought, CAGRD staved off a supply crunch by making a deal with GRIC in 2018. But buying CAP water, either directly or through a third party, like GRIC, has become even more difficult after the mandated cuts. The rickety practice of paying now for water that might be replaced in the future is wearing thin, evoking the land frauds that have stained Arizona's history. As Jake Bittle argues, a generous interpretation of CAGRD is that it acts as "a balance sheet for the region's total water flow" to ensure there is no overall net deficit, but the reality is that it was conceived as "a giant loophole in the state's groundwater restrictions" that "deferred a difficult reckoning to someday in the distant future."[35] In the meantime, CAGRD has effectively become the region's developer by default.

Encouraged by the ongoing softening of the GMA regulations, predatory Wall Street investors made their move to buy up rural land, close to the Colorado River, with a view to selling the water to thirsty communities closer to Phoenix. One company, Green-stone Management Partners, bought nearly 500 acres of land, and promptly flipped the water rights for $27 million to Queen Creek, an

---

34 Kathleen Ferris and Sara Porter, "The Elusive Concept of an Assured Water Supply: The Role of CAGRD and Replenishment," Kyl Center for Water Policy, Morrison Institute (Phoenix: Arizona State University Press, 2019).
35 Jake Bittle, *The Great Displacement: Climate Change and the Next American Migration* (New York: Simon & Schuster, 2023), 200.

East Valley suburb almost two hundred miles away. The founder of another hedge fund, Water Asset Management (WAM), described Western water as a "trillion-dollar market opportunity," making it the "biggest emerging market on Earth." WAM closed several deals in Arizona and other Western states, while hiding its name under a variety of LLCs; they were all traced, by CNN investigative reporters, to the firm's HQ in Manhattan.[36] These East Coast speculators are drought profiteers, grabbing land for its subterranean assets. But the state's complex market of inter-basin transfers had been fostering "water farms" like these for decades, even though the establishment of CAGRD was supposed to do away with them.

Another of WAM's likely clients is the extremely thirsty West Valley city of Buckeye, thirty miles from downtown Phoenix, which has mushroomed from a small agricultural town to one of the fastest growing boomburgs in the country. Buckeye now spans 640 square miles, larger in area than Phoenix itself, and has approved twenty-seven master-planned communities, including Teravalis, at fifty-seven square miles, the country's largest, currently mired in a lawsuit over whether it has enough groundwater approvals to commence building. With only 108,000 residents in 2024, some of Buckeye's local boosters were zealously imagining the population growing to 1.5 million. It does not look or feel like any city I have visited. There is a downtown pocket, branded as "Historic Buckeye," consisting of a one-stop retail cluster, featuring a gun supply store, two taquerias, a Dollar General, a liquidation center, a Western wear outlet, rodeo restaurant, and drive-through cigar shop. From there, you can walk out a short distance to active farmlands, while Buckeye's low-density residential subdivisions are far-flung, scattered all over the land between the Gila River and the White Tank Mountains. These colonies are interspersed with large row acreages of alfalfa, sorghum, winter wheat, and corn. Agribusiness processing plants and warehouses are the biggest structures, lodged on the corners of rural sections. There is no urbanism to speak of in Buckeye—whatever community energy it has is centrifugal, dispersed, and scattershot.

---

36 Lucy Kasanov, "Wall Street is Thirsty for Its Next Big Investment Opportunity," *CNN*, March 22, 2023, https://www.cnn.com.

So far, the city has relied for its growth on groundwater from the now overdrawn Hassayampa Sub-Basin. Of all the suburban communities affected by the governor's 2023 moratorium, Buckeye was hit the hardest. Its officials described the order as a "reactionary response" to an "outdated model" from an earlier economic era. In desperation, they turned to water-hunters who scour the region to find new sources, but the market price for new water is soaring. City Hall ended up forking over a cool $80 million for a single, water-rich acre of land in the Harquahala Valley,[37] a basin from which the city was permitted to transport water. In the meantime, officials were talking about "pie in the sky" plans, such as building a desalination plant next to the Sea of Cortez to treat seawater and then piping it several hundred miles north, and uphill, to the West Valley.

Builders were sitting on thousands of unbuilt Buckeye homes already certified through CAGRD, but there were countless others that were technically "stranded" by the moratorium. As always, industry pressure on elected officials produced results. The governor convened a Water Policy Council to propose a rescue package called Alternative Designation of Assured Water Supply (ADAWS).[38] Under this workaround, the state would allow for fast-growth cities on the fringe, like Buckeye, to be "designated" as water providers, making the municipality, rather than the developer, responsible for assuring supplies of renewable water. The program would also "provide a path for closing statutory loopholes associated with 'wildcat' subdivisions and 'Build-To-Rent' developments." The governor's council considered several ways for municipalities like Buckeye and Queen Creek to add to their water portfolios without mining more groundwater. Sara Porter, the director of Arizona State University (ASU)'s Kyl Center for Water Policy, who served on the council, explained to me that one of the problems was:

---

37 Amy Scott, "The $80 Million Acre," *NPR Marketplace*, October 2023, https:// www.marketplace.org.

38 Jeremy Duda, "Hobbs Pitches Water Plan to Resolve Housing Moratorium That Hit Buckeye and Queen Creek," *Axios*, January 16, 2024, https:// www.axios.com.

[The] places where home builders and developers want to go won't necessarily fall within these designations. [They] desire to do master-planned subdivisions in undeveloped locations— with mountains and foothills as vistas, and the remaining farm- lands are not in a desirable location. . . . They are making the choice not to build on alfalfa fields, and choosing to develop in more spectacular parcels, where the aquifer is not so deep, and where wells can go dry.

Incentives would have to be offered, Porter pointed out, to persuade "developers to buy and retire farmland and use a little of that ground- water already being used by growers for housing." Terry Lowe, the city's water resources manager, had other explanations at hand. "The groundwater closer to downtown is brackish," he told me, "and of a poor quality regarding potability, which means it's OK for alfalfa, or other feedstock crops, and is easy to pump, but developers want the better quality of water they can find further out."

If the ADAWS plan was implemented to ease the path of growth in Buckeye, Porter believed that any new users of ground- water should be enrolled in CAGRD and that "its replenishment should occur within hydrological impact areas." By the end of 2023, 383,000 homes were enrolled in CAGRD. At its inception in 1993, CAGRD's directors assumed that excess CAP water would be avail- able for replenishment, but that surplus had long since dissolved. There are physical limits to what can be pumped from the aquifers before wells run dry, or lowered water tables led to land subsidence; fissures open up in the ground, as has occurred in Pinal County, and the aquifer collapses. Faced with the prospect of further CAP cuts, and the dismal math that shows a groundwater deficit in many of the areas slated for residential development in the Sun Corridor and the West Valley, water attorneys and brokers intimately connected to the housing industry are pressuring officials to make good on their promises of "augmentation." This kind of magical thinking assumes that, in the near future, a combination of technological ingenuity and political savvy will summon forth a new source of surface water. The current contenders include a ramping up of recycled effluent, expansion of the reservoir above the Bartlett Dam in the Verde River, a renewed effort to acquire Colorado River water from recalcitrant on-river users, more Indian leases, and the last, albeit ultra-expensive, resort of desalination.

In Buckeye, Lowe's job was to go out and secure any or all of the above for the city's portfolio. He added that he was eying local farmers who had surface water rights that they might sell to the city. And then, he said, "there is always the Next Big Thing." CAP had been the last major augmentation project, and now its future sustainability was in doubt. Lowe did not consider the Mexico desalination plan to be "too far-fetched." Expensive yes, but "people thought $4b for CAP would be too expensive." The state, he pointed out, "has been good at water management and would find a way." In the meantime, he would continue to play the role akin to that of a water diviner for his municipal employer.

SUN CORRIDOR BLUES

When it comes to population statistics, Maricopa County has dominated the charts for decades as the nation's largest-gaining county; it added almost 57,000 new residents in 2022 to the 4.5 million already living there. But the less populous Pinal County, the rural county to the southeast of Phoenix, is the state's fastest growing overall, and boosters were fond of proclaiming that three of its cities, Queen Creek, Casa Grande and Maricopa, were among the nation's ten fastest-growing cities with populations of 50,000 or more. Casa Grande even attracted a large manufacturing facility—a rare kind of business investor in a region where cities compete for water-guzzling data centers. The company, Lucid Pure Air, started moving electric vehicles off its robot-intensive assembly lines in 2021.

Pinal spans the hundred miles from Phoenix's suburban fringes to those of Tucson. Slated to connect the two cities at the center of the "Sun Corridor" megaregion, the county is the apple of the housing industry's eye—cheap farmland waiting to be filled with brown-tiled roofs. The Sun Corridor, running from Prescott to the border city of Nogales, was dreamed up by Arizona's boosters as a branding opportunity that also suggested a bonanza for the real estate industry. There was frothy talk about trillions of dollars in private and public investment, and in Pinal alone, 650,000 residential housing lots were entitled in anticipation of subdivisions breaking ground.

This promised land has presented some real challenges. Pinal occupies the number two spot on a list of the country's counties most at risk of becoming uninhabitable in the next twenty to forty years due to climate change.[39] In 2021, after a hydrological model run revealed 8 million acre-feet in unmet demand over 100 years, the Department of Water Resources stopped approving applications for new assured water supply certifications for future subdivisions in the county. As with the governor's 2023 order to halt new approvals in the Phoenix suburbs, lots already approved were not affected and so building will not cease.[40] But the writing was on the wall for anyone still dreaming of pell-mell development in the corridor.

The list of disappointed stakeholders included farmers whose business model in Central Arizona is to grow until the bulldozers and cement mixers are in sight and then to retire on the proceeds of a land sale. The growers were already suffering from stricter regulations on groundwater usage and the mandatory reductions in CAP allocations. Pinal's farmers, who use 90 percent of the county's water, took the biggest hits, and in January 2023, were cut off entirely from CAP's surface water. They are among the nation's top producers of commodity crops like alfalfa, cotton, barley, and durum wheat. In particular, alfalfa requires copious amounts of water, but there is high local demand for the crop from dairy and beef producers. This perennially flowering plant, which produces high-protein hay for cattle feed, is so sought after that a large Saudi Arabian company bought 10,000 acres of Arizona land in 2014 to supply its dairy cows, after its local Saudi groundwater supply ran dry. Al Dhara Farms, a UAE company, followed suit.[41] Government subsidies keep these water-guzzling acreages in business, but, with the rationing of surface water, much of

39 Al Shaw et al., "New Climate Maps Show a Transformed United States," *ProPublica*, September 15, 2020, https://propublica.org/climate-migration/.

40 See the report by ASU's Water Resources Research Center, "Getting Down to Facts: A Visual Guide to Water in the Pinal Active Management Area," June 12, 2020, https://wrrc.arizona.edu.

41 Coverage of the Saudi arrangement generated such a public backlash that even Bruce Babbitt, the former governor and dean of Arizona's water politics, weighed in against the lease. "Hobbs Should Terminate the Saudi Lease in Butler Valley," *Arizona Capitol Times*, July 28, 2023, https://azcapitoltimes.com.

Pinal's irrigated 200,000 acres of arable land has been lying fallow or running off rapidly depleting groundwater.[42] If growers can't afford to pump from deeper aquifer levels or make deals for tribal water, they will not be able to plant, and if the brown-tile roofs don't come to bail them out, their migration from the land is assured, though no one, so far, is anticipating the kind of mass exodus that emptied out the Great Plains during the Dust Bowl years.

The Sun Corridor concept was launched on the back of an enthusiastic 2008 report, *Megapolitan: Arizona's Sun Corridor*, from ASU's Morrison Institute for Public Policy.[43] According to the study, whose lead author was the well-known developer attorney Grady Gammage Jr. (best described as Arizona's preeminent real estate intellectual), nearly 8 million people would be living in the 32,000 square-mile region by 2030, an 82 percent increase over the 2000 population. Without doubt, the jewel in the Sun Corridor crown was the 276 square-mile plan for Superstition Vistas (SV), a new city, twice the size of Tucson, nestled between the urban edge of Apache Junction and the iconic Superstition Mountains. To be built on raw desert land, currently owned by the state, the SV plan envisaged housing as many as a million people by its 2090 buildout, and was touted as a model of sustainable development on a grand scale.[44] The ASU's Mor-

---

42 Nina Lakhani, "The Farmers Dealing with Water Shortages Even Before Historic Colorado River Deal," *The Guardian*, May 31, 2023, https://www.theguardian.com.

43 Grady Gammage Jr. et al., *Megapolitan: Arizona's Sun Corridor*, Morrison Institute for Public Policy, Arizona State University (May 2008), https://morrisoninstitute.asu.edu.

44 After the housing crash brought on an economic downturn, Gammage's follow-up report, *Watering the Sun Corridor: Managing Choices in Arizona's Megapolitan Area* (August 2011), revised the population estimate to 9 million by 2040, and expressed more caution about water availability for all this growth. See https://morrisoninstitute.asu.edu. In his book, *The Future of the Suburban City: Lessons from Sustaining Phoenix* (Washington, DC: Island Press, 2016), Gammage was still forecasting a Sun Corridor population of 9 million if all agricultural use were eliminated, and with per capita conservation, a population of 12 million. In 2021, after the drought had deepened, concern about water availability was even more pronounced when he revisited the topic, in "Return to Watering the Sun Corridor: A Perspective by Grady Gammage Jr.," Kyl Center for Water Policy, Spring 2021, https://morrisoninstitute.asu.edu.

rison Institute also bestowed its imprimatur on the project, thereby lending a public-interest gloss to one of the most grandiose dreams of Arizona's growth machine.[45] Promoted just as the downturn from the financial crash was about to curve into a near-death spiral for the housing industry, the Sun Corridor/SV package was the last, adrenaline-rich spurt of the fast-growth years, and it came with progressive bona fides.[46] SV's advocates summoned up all the bells and whistles of thoughtful urban planning; knowledge economy jobs, light-rail networks, solar power, and higher-density, mixed-use zoning. State ownership of the land, to be sold carefully to developers, promised at least a degree of control over the plan's buildout, in contrast to the usual piecemeal, leapfrog patterns of development. If planned properly, the state could even be a co-developer, ensuring a more sustainable pattern of building.

There was, at least, one big problem. SV had no designated water source, and, like everyone else in the Sun Corridor, it would have to compete with a multitude of developers and municipalities in the region for access to new supplies.[47] This was a costly prospect in the 2000s when the SV concept was launched, but even more so by 2021 when ground was finally broken. As housing prices recovered after 2014, and Greater Phoenix regained its evergreen status as a hot market, the state saw its opportunity to get a high return on piecemeal SV land sales. In November 2020, the Arizona State Land department auctioned off SV's first parcel—2,783 acres of desert scrub south of Apache Junction—for $245.5 million, almost $177 million above the market appraisal of $68 million.[48] Yet that same year, the hydrological model run by another state agency, the ADWR, showed that Pinal County's aquifers could not guarantee a long-term supply

45 Grady Gammage Jr. et al., *The Treasure of the Superstitions: Scenarios for the Future of Superstition Vistas*, Morrison Institute for Public Policy, April 2006, https://morrisoninstitute.asu.edu.

46 Jonathan Thompson, "A Dizzying Look Back from Phoenix's Future," *High Country News*, June 15, 2023, https://www.hcn.org.

47 William Pitts, "Newest Arizona City Already Facing Major Water Problem Before It's Even Built," *NBC Channel 12*, May 10, 2022, https://www.12news.com.

48 Lisa Honebrink, "Land Zo: What Does the Sale of Controversial State Trust Land Portend?" *Phoenix Magazine*, January 15, 2021, https://www.phoenixmag.com.

for building at the current rate. Water was negotiated—for the four square miles of SV land on sale, from Apache's Junction's assured supply—but there was no guarantee that further rights could be secured.

Nor did the identity of the buyer at auction augur well for SV's first phase. The winning bidder (in conjunction with Brookfield, the Canadian real estate mega-developer) was D.R. Horton, the nation's largest homebuilder by volume. Among the big merchant builders, Horton was best known for spitting out the kind of mono-cropped subdivisions that are the Satanic target of anti-sprawl advocates. It was an odd match for a development with the patina of "enlightened real estate," previously backed by university sources and respected nonprofit organizations like the Sonoran Institute and the Lincoln Institute of Land Policy. Horton's price points for its first Radiance subdivision began at $403,000, while the average home price in adjacent Apache Junction was $377,000. Forty miles from downtown Phoenix, and without any guarantee of enjoying the New Urbanist features depicted in SV's initial plan, buyers would be paying a stiff price for living on the very edge.

On my visit to the Radiance sales center, I drove through the hopscotch subdivisions of Queen Creek to the south, and up Ironwood Road toward the famous Superstition Mountain range. Like Buckeye, Queen Creek was malingering under the governor's moratorium, with many stranded lots. En route, I passed a lot that had just been sold by the Church of Jesus of Latter-Day Saints (LDS), one of Arizona's largest landowners, for a build-to-rent subdivision, which was currently exempt from the water restrictions. It's obvious when you reach the outer boundary of metro Phoenix; there are brown-tiled roofs on one side of the road, and scrubland on the other. At this razor edge, clusters of large yellow machines are busy moving earth around, and America's national merchant builders—Pulte, Lennar, Meritage, KB, Hovnanian, Horton, Taylor Morrison—will greet you with signs and fluttering flags promising "New Homes at Reasonable Prices," "Now Selling – Great Neighborhoods in a Resort Setting," and urging you to "Join the Waitlist." Out here, the prices are supposed to be lower—in line with Central Arizona's mantra of "drive until you qualify"—but the spectacular Superstition Mountain views, and easy access to the range's parks and hikes, are a premium for would-be residents of highly promoted developments like Radiance.

The ethereal name, and the strenuous branding, of Radiance suggests an upscale master-planned community, but I found that the first phase, with four hundred people already in residence, to be more than a little humdrum, with low-walled compounds of compact, beige one-and two-story units, hosting a limited range of models, and standard curving roads. "We only do single family homes, we really like them," gushed a sales associate, adding that "we are bringing the spruce to Apache Junction," known for its goldmining-era ghost town. Swimming pools, schools, and stores would follow, she assured me, but so far, the reality of SV's first chapter looked exactly like business as usual for a housing development on the urban fringe. Further up the Apache Trail highway, in a hiking trail parking lot, I met a sun-weathered woman, walking a cat on a leash, who had lived locally all her life and was now an RV itinerant. This lifestyle was not by choice. "All the new folks from New York, California, and Canada have gone and priced me out," she explained, "and the homebuilders have taken all the water." She said she could never leave the area—"these mountains have a hold on me"—but had gotten used to moving her trailer around, except in the summer, "when my flip flops melt and stick to the road."

PHOENICIANS

The cheapest homes in the region—in South Phoenix's poverty tracts—are well supplied with water from the Salt River distribution system, but their occupants have long been beset by other environmental challenges, including toxic industries on their doorstep. Historically, Phoenix's racialized minorities have been segregated and consigned to below the Salt's old floodplains. The area once boasted the dirtiest zip code in the country and was regularly cited as a textbook example of environmental injustice. Climate change has added a new dimension. Phoenix is already one of the world's hottest cities, surpassed, at the height of the summer months, by only a handful of Middle Eastern locations.

As the blistering heat builds through July and August, when overnight lows can linger in the nineties (hitting a record of 97°F one night in July 2023), it takes a disproportionate toll on residents of low-lying ground where there is insufficient shade to deflect the sun's

rays. According to a city report, Central Phoenix, with high propor-
tions of Black and Brown residents, has just 6.1 percent canopy coverage.[49] The districts with the least tree cover are south of the river
and the railway tracks, where neighborhoods are 10 degrees hotter
than those in others parts of Maricopa County, according to American Forests' Tree Equity Score.[50] Nor is the abundant sunshine being
harvested in a productive way. Largely because of political resistance
from GOP lawmakers and the obstruction of its largest utility company, Arizona only draws 10 percent of its energy from solar sources.

With relatively few hands to play against the most palpable
impact of global warming, Mayor Gallego appointed ASU professor
David Hondula to head up a new Office of Heat Response & Mitigation in 2021, and launched a Tree and Shade Program, boosted by
federal dollars, to plant drought-resistant trees like elms, ashes, and
Chinese pistaches.[51] The city also rolled out a sister program, called
Cool Pavement, that sealed hundreds of miles of roadways and sidewalks with a coating that reflects, rather than absorbs, heat from the
sun. Officials claimed that the result is a ten to twelve degree difference in surface temperatures during the midday hours, although a
subsequent study showed that the reflected heat from those surfaces
makes pedestrians warmer by more than five degrees.[52] Among the
green gizmos developed within ASU's School of Sustainability is the
first mechanical tree, one thousand times more efficient at mitigating
heat than a real one, though, of course, it produces no oxygen.[53]

City Hall and university researchers have technically sweet programs and gizmos on offer, but it is the city's community activists,
like Matthew Perea, director of Chispa (a program of the League of

49 "Tree Canopy Cover," City of Phoenix, https://www.phoenix.gov.

50 "Tree Equity Score," *American Forests*, https://www.americanforests.org.

51 Serena O'Sullivan, "New $10 Million Tree Planting Grant Will Expand
Phoenix Shade Canopy," *Phoenix New Times*, October 23, 2023, https://www.
phoenixnewtimes.com.

52 Amy Cutler and Cody Lillich, "Phoenix Expands 'Cool Pavement,' Research
Shows It Makes People Hotter," *Arizona's Family*, July 13, 2023, https://www.
azfamily.com.

53 Elias Weiss, "World's First Mechanical Tree 'Planted' in Tempe, and It
Sucks. In a Good Way," *Phoenix New Times*, April 19, 2022, https://www.
phoenixnewtimes.com.

Conservation Voters), who have firsthand experience of how uneven the impacts are from neighborhood to neighborhood. As Perea put it: "the current mayor believes in climate change, but not climate justice." A construction worker and union organizer for the Roofers Union during the years of rapid housing growth, Perea knows how arduous it is to labor in high temperatures. Unlike in Gulf countries, where construction is supposed to halt from 12:30 p.m. to 3 p.m. during the summer months, there are no such protective regulations in Arizona.[54] Nor, according to him, are many workers' deaths recorded as part of the city's official statistics on heat-related fatalities. As temperatures rise across the US, OSHA protections lag far behind.

Perea, who emigrated from Chihuahua in 1995, lived for a long time on the low-lying West Side of Phoenix, where the heat island effect is almost as pronounced as in South Phoenix—as much as ten degrees hotter than in the more affluent North and East—and so, he welcomes the campaigns to triple the tree canopy in both of these neighborhoods. National surveys of "tree equity" have shown a clear disparity in tree cover between affluent neighborhoods and those with predominantly Black and Brown populations still dealing with the legacy of redlining. The new saplings that dotted the previously bare avenues and walkways of South and West Phoenix did not look like much of a match for the summer heat, but, over time, they will make a difference, at least if they are well maintained. In that respect, Perea cited Chispa's vision for trade schools, where community members could be trained for green jobs in arboriculture, to prune and tend to the new trees, and he envisaged worker coops specializing in these kinds of green service occupations. He also insisted that environmental justice had to be more broadly understood; community health issues like police brutality, free public transport, affordable housing, and food deserts were just as "green" as planting trees and repaving roads and sidewalks in low-income neighborhoods. "We

---

54 OSHA regulations hold that workplaces with dangerous levels of heat stress can be subject to penalties of only up to $16,000. In July 2024, the Biden administration proposed stricter rules for workplaces when the heat index reaches 80°F or above, including increased access to water and temperature-controlled break rooms. Dharna Noor, "Biden Attacks Republican Climate Deniers as He Unveils Extreme-Heat Rules," *The Guardian*, July 2, 2024, https://www.theguardian.com.

have to change the narrative," he insisted, "not just at City Hall, but in the communities themselves."

Chispa was a co-founder of the Phoenix Environmental Justice Alliance, along with Black Lives Matter (BLM). The formation of this alliance was the fruit of an uneasy relationship between active Black and Brown groups in the city. BLM's Yeshaq Tekola, a whirlwind of movement energy whom I met later that day, shared Perea's view that environmental justice was an all-encompassing outlook; "there's not much that lies outside of it," they insisted. As abolitionists, their BLM core of "queer Black neurodivergent feminists" were poles apart from the city's older, and more culturally conservative, civil rights organizers who had cut their teeth on the classic paradigm of environmental injustice: toxic facilities built next to African American housing. As a younger generational cohort weaned on horizontal, or nonhierarchical, organizing, Tekola's circle took their cue from community accountability rather than looking to elected officials for policy fixes. Hailing from Seattle's bastion of sustainability, Tekola seemed to relish the opportunity to start from scratch in Phoenix, pointing out that "here it is easier to build the table you want to sit at." As abolitionists in Arizona, they had to consider the question: "When the taps go dry, how will the police respond?" Knowing full well that protection will not be offered on the basis of equality, their long-range vision had to be "a world without police, prisons, and borders." It was a far cry from the world where they lived, where frequent bomb threats from white supremacists meant that the BLM office was unmarked from the outside, even in a historically Black neighborhood.

The late February day that I met with Tekola, Phoenix's updated Heat Response Plan was submitted to City Council. After a prolonged email exchange with a variety of City Hall employees and officials it was decided that no one would meet me to talk about it. With spring in the air, and highs already touching 80 degrees, everyone seemed nervous about the warmer months that lay ahead. The previous July, temperatures reached 110°F or above thirty times, and 115°F or above seventeen times, while the seasonal monsoon was the driest since 1895. The national press feasted on the grisly spectacle of a major American city transformed into a blast furnace. The summer of 2023's death count was a source of great embarrassment to City Hall, and over the course of the year—globally the hottest ever recorded—Maricopa County officially counted more than 645 heat-related deaths.[55]

Stacey Champion, an environmental consultant to the state attorney general who had previously served on the previous mayor's sustainability council, had been scrupulously gathering heat mortality data from the medical examiner's office for years. Indefatigable in her efforts, she had won a long battle over her right to lodge public requests on the matter. "If there were a PhD for heat-death studies, I would have it by now," she declared. "The deaths begin as early as April," according to Champion, "and people keep dying until October." When asked what would make for a successful policy approach, she was quite blunt: "Care more about poor people and save them from dying." As she pointed out, "if you go from one air-conditioned environment to another, you are simply not aware of what it's like for people on the street or who can't afford to keep their units running at a safe temperature. If the A/C unit of an elderly person living alone happens to break down in the height of the summer, they are in real trouble here in Phoenix." According to one study, if a general power outage were to occur, disabling air-conditioning throughout the metro area, nearly 800,000 people would require emergency health care, and as many as 13,000 residents would die.[56] A/C units already account for about one-fifth of all energy used from buildings, and, in Sunbelt climates, they shield us from the stark reality of a warming world, as well as from older ways of coping with the heat. As Jeff Goodell points out in his grimly titled book *The Heat Will Kill You First*: "Air-conditioning is not just a technology of personal comfort; it is also a technology of forgetting."[57]

---

55 Nationally, according to the EPA, extreme heat was linked to some 11,000 deaths and 120,000 emergency room visits last year. "Climate Change Indicators: Heat-Related Deaths," Environmental Protection Agency, https://www.epa.gov.
56 Brian Stone Jr et al., "How Blackouts during Heat Waves Amplify Mortality and Morbidity Risk," *Environmental Science and Technology* 57, no. 22 (2023): 8245–55.
57 Goodell reported that he decided to write his book after nearly passing out during a fifteen-block walk in downtown Phoenix in high summer. Jeff Goodell, *The Heat Will Kill You First: Life and Death on a Scorched Planet* (New York: Little, Brown and Company, 2023).

Every study of heat-health impacts shows that stress and deaths are much higher in lower-income communities of color, in houses with substandard, poor insulation and older A/C units. These impacts are particularly brutal among people experiencing homelessness, who account for roughly half of heat-related fatalities in Maricopa County.[58] There are few sharper illustrations of environmental injustice than the uneven effects of a sustained heatwave, where the poor and unhoused suffer while the well-heeled are insulated by artificially cooled air. The plight of migrants struggling to cross the searing borderland deserts attracts close attention and care from Tucson's humanitarian groups.[59] Less well-acknowledged are the urban homeless—a large population in Phoenix. From 2016 onwards, this population has been concentrated in the sprawling encampments of the Zone, stretching for fifteen blocks of downtown Phoenix.

Toward the end of 2023, their tents, handcrafted shelters, and tarps were summarily cleared by officers of the Phoenix Police Department who earned an unprecedented rebuke by the US Department of Justice (DOJ), in June 2024, for their long-standing record of brutality and racial discrimination.[60] Champion had done volunteer outreach in the Zone for several years. During the pandemic, when Phoenix police moved the encampments into vacant spaces with no shade, she called them out publicly as the "death lots." She was often threatened with arrest for trespassing, once for passing out Gatorade, and on another occasion, was cited for trying to prevent officers from destroying the property of a homeless person. Because of her humanitarian presence on the streets, she was able to provide a decade's worth of key documentation to DOJ investigators about the egregious police harassment of its unsheltered residents, along

---

58 Osha Davidson, "For Unhoused People in America's Hottest Large City, Heat Waves Are a Merciless Killer," *Yale Climate Connections*, July 24, 2023, https://yaleclimateconnections.org.

59 The *Undocumented Migration Project*, directed by visual anthropologist Jason De León, has documented this suffering under intolerable heat conditions. See https://www.undocumentedmigrationproject.org.

60 "Justice Department Finds Civil Rights Violations by Phoenix Police Department and City of Phoenix," US Department of Justice (June 13, 2024), https://www.justice.gov.

with seizure and destruction of their property. The statistics cited by the DOJ's charge of "very significant and severe violations of federal law and the Constitution" showed that people experiencing home-lessness made up 37 percent of the misdemeanor arrests and citations between 2016 and 2022 while making up less than 1 percent of the city's residents.[61]

When I returned to Phoenix in mid-June, just days after the DOJ report made the headlines, daytime highs were already hitting 110 degrees. The intolerable heat had started earlier this year, as predicted, and the metro region was looking at a prolonged period of above-110°F temperatures. In fact, the summer ended up being the hottest on record with average temperature over June, July, and August at ninety-nine degrees, and zero days with highs below 100. Champion told me she was still receiving files of heat fatalities from the Medical Examiner's office from last July. The methodology for analyzing deaths had changed, and so there were considerably more deaths this year already, registered on the county's gruesomely named "Death Dashboard." By the end of the month, the fatalities were up 84 percent from the year before. City Hall had set up more than forty cooling centers and several respite centers across the city, as part of a regional Heat Relief Network which comprised more than two hundred facilities. Relief could be sought there during the afternoon hours and, in some centers, in the evenings too.

Before we set out to visit some of them, Champion and I noticed that the app for the network map was broken. Asking around, we learned that many were not open on a Sunday and only one respite center was accessible overnight. It was basically a room with A/C in the annex of the city's central library. There was nowhere to rest your head except for a few tabletops. People were trying to sleep in the park at the back of the library, but they told us that police had been moving them on by threatening them with trespass arrests. In another example of awful timing, city council had just passed a ban on camping within five hundred feet of Phoenix's public parks.

61 Madeleine Parrish and Nick Sullivan, "5 Key Takeaways from the Justice Department Report on Policing in Phoenix," *Arizona Republic*, June 14, 2024, https://www.azcentral.com.

With all the city's shelters full, the only other after-hours respite was a "structured campground" called SOS (Safe Outdoor Space), where tents were closely packed together under an outdoor canopy in the former Zone. When Champion and I tried to visit we were gruffly warned off by private security guards. From a distance, it resembled a detention camp, and the guards' hostile attitude only reinforced that impression. By the time we got to the fourth stop on our tour, the heat had gotten to me; I had a headache and felt dizzy. To make matters worse, a breeze sprang up, doing its best to mimic an assault by a hairdryer at maximum setting. For pedestrians, official air temperatures can be misleading because they do not record radiant heat. In an urban environment, at 110 degrees, the load on the human body can be much higher—more like 130 degrees—because of all the heat being radiated from surfaces.[62] Another gauge, often used to calculate heat stress, is the wet-bulb temperature, which considers wind speed, relative humidity, cloud cover, and sun angle. I had only been outside for a few hours, mostly in the shade, and needed respite. It was difficult to imagine people on the street enduring these conditions week after week without incurring severe bodily harm. Like other cities, Phoenix had tried to deconcentrate and disperse its homeless populations: "No Trespassing" signs were posted all over the blocks of the former Zone, which also featured bollards to prevent camping. It seemed as if the city was letting the summer heat take care of its "homeless problem"—by killing the people it would not house.

Homelessness jumped after rents began to rise in 2016. These price hikes were, in part, an unintended consequence of the successful push from anti-sprawl advocates of urban sustainability to build higher-density housing on infill areas in center cities. Downtown Phoenix, which, until then, had been relatively affordable for low-income residents, became an eligible location for young, college-educated workers with resources. Another impact has been the

---

62 ASU researchers have developed a robot, called MaRTy, to determine "mean radiant temperature," as experienced by pedestrians. Robin Tricoles, "Street Smarts Required in Heat Mitigation," *ASU News*, May 5, 2020, https:// news.asu.edu.

addition of acres of heat-absorbing hardscape. In Sunbelt cities, this aggregation of surface area has amplified the urban heat island effect. Formerly hollowed-out downtowns in places like Phoenix (where ASU has constructed a large campus with superblock buildings over the last decade) are growing their surfaces vertically and are getting hotter as a result. Downtown Phoenix had filled up quite noticeably since I was last there. But the new multifamily housing, built on dirt lots that were vacant for decades, has also come at a cost for renters, or would-be owners. Prices have risen at a higher clip than anywhere else in the Valley. By May 2024, the median downtown home price had risen to $625,000 ($100,000 above the overall figure for the city of Phoenix), and it had become the most expensive metro neighborhood for rentals, with unit rentals climbing above $2,000 a month. According to the National Low Income Housing Coalition, Arizona was second only to Nevada and California in its shortage of available and affordable housing for low-income dwellers.[63] To afford housing in the city, you would need to be earning at least $25 an hour, far above the minimum wage.[64]

Like other cities hard-hit by a shortage of affordable housing, Phoenix launched its first-ever Housing Plan in 2020, with the goal of creating or preserving 50,000 homes in the course of the next decade. Over the years, the city had banked many land parcels. In 2016, it owned an estimated ninety square miles of land, of which 2.3 square miles was entirely vacant.[65] As a result, it ought to have been able to build more cheaply under the 2020 plan. Given the entrenched community opposition to multifamily housing, and the difficulty of re-zoning, it proved to be tough going, but several projects, all undertaken by private developers, were being approved and built. One

---

63 The National Low Income Housing Coalition, "Arizona," *The GAP* (2023), https://nlihc.org/gap/state/az.

64 The National Low Income Housing Coalition, "How Much Do You Need to Earn to Afford a Modest Apartment in Your State?" *Out of Reach*, https://nlihc.org/oor.

65 Dustin Gardiner, "'Bad Neighbor': Phoenix Struggles to Manage its Vacant City-Owned Lots," *Arizona Republic*, November 24, 2016, https://www.azcentral.com.

gauge of whether the city was succeeding in this endeavor was the eviction rate. In 2023, according to Princeton's Eviction Lab, there were more than 84,000 eviction filings, or 32 percent above average, and by January 2024, the rate had risen to 41 percent above average.[66] Many of the residents turned out of their homes ended up in shelters or in the streets, facing down the brutal summer heat. Catastrophic weather, including extreme heat, also undermines the integrity of homes themselves, degrading roofs and walls, and causing foundations to shift. In this way, it contributes to the instability of available housing stock, especially cheaper units that are then torn down and replaced by less attainable ones. In regions like the Southwest, in the bull's eye of global warming, it is more and more apparent that housing is not just a challenge for migrants displaced by flood, drought, or food insecurity, but also a domestic climate justice issue.

ARIZONA'S ISRAEL

American identification with Israel draws on much more than a shared history of settler colonialism. The dominant ethnic groups in both nations claim providential origins and see themselves through the biblical lens of a people chosen by God to occupy a promised land. In particular, the Mormon migration and settlement of the interior West was one of the most strenuous missions of Christian Zionism. Joseph Smith's promised land of Deseret had much more in common, topographically, with historic Palestine than the New England forests colonized by Puritans who were bent on building their own new Jerusalem there. The arid lands of the West were soon being imagined as akin to a Middle Eastern homeland. Mormon pioneers migrated to the Arizona territory from Utah, acquiring and irrigating large tracts of land, and eventually occupying a key role in state politics. They also carried with them a doctrinal belief that they were Israelites, of the house of Jacob, and that biblical figures were their direct ancestors.

---

66 "Phoenix, Arizona," Eviction Lab, Princeton University, March 2024, https://evictionlab.org/eviction-tracking/phoenix-az/.

During my first return visit to Phoenix, I stayed in Mesa, an expansive suburban satellite with 500,000 residents, settled by Mormons in 1878, and often named as America's most conservative city. Mesa's founding fathers labored to refurbish a Hohokam ditch and to bring Salt River water to the location chosen for settlement. The colony was surveyed "with a straight edge and a spirit level," and platted according to the "city of Zion" plan—a rigid-grid staple of Mormon town design. Today, its heritage museum and Pioneer Park pay homage to the feats of reclamation and sacrifice that accompanied the original "conquest" of the land and its Indigenous occupants who would be assigned a significant rank in the LDS hierarchy. The Mormon residents I spoke with expressed awe at the hardships of their irrigationist forebears. A young woman who staffed the LDS Temple's visitor center talked fervently about her own personal connection to ancient Israel: "our prophets were from one of the tribes of Israel," she told me, "and so I am related to them, either by adoption or genealogically." Another explained that "bringing water into the desert was something that the Israelites did before us, and they continue to inspire us thousands of years later." To illustrate her point, she cited a verse from the Old Testament (Isaiah 35:1–2): "The wilderness and the solitary place shall be glad for them; and the desert shall rejoice, and blossom as the rose."

The Mormons' irrigationist feats in the Great Basin and in Southwestern settlements like Mesa laid the basis for their new Zion. But they also fueled the hucksters' bogus promotions of Western settlement "beyond the 100th meridian," despite the warning of John Wesley Powell that agrarianism was not sustainable in these dry lands. Indeed, for many years, as Reisner noted, the Bureau of Reclamation was "based on Mormon experience, guided by Mormon laws, and run largely by Mormons." The LDS successes in watering and "taming the wilderness" of the West would also be echoed thousands of miles away in Palestine where Zionist settlement in the 1920s and 1930s hinged directly on water diversion and irrigationism to "make the desert bloom." In both locations, settlers adopted the mantle of a civilized people destined to succeed the backward natives as the natural inheritors of the land, along with a demographic mentality obsessed with the need to maintain a numerical superiority of white people. As an early twentieth-century promoter put it, Phoenix was "a modern town of 40,000 people, and the best kind of people too. A very small percentage of Mexicans, negroes, or foreigners."[67]

Over time, the material ties between these two climatically kindred regions evolved, through exchanges of knowledge about arid-land agriculture to desalination technology. Arizona's key role in the regional "Gunbelt" of military production during the Cold War years positioned it well to partner in Israel's military build-up in the decades after its Occupation of the West Bank and Gaza commenced in 1967. Israel's Iron Dome anti-missile defense system was developed in Tucson by Raytheon, while Elbit, Israel's premier arms manufacturer, provided much of the security infrastructure for the militarization of Arizona's borderlands in the years after 9/11. The state itself was a prime beneficiary of US aid to Israel, most of which took the form of contracts to domestic defense firms. This highly profitable relationship, already flourishing through the Arizona Israel Technology Alliance, was boosted in 2019 when Arizona opened its own trade and investment office in Tel Aviv to facilitate business ventures between Arizona and Israeli firms. Tucson had its own local version, the Israeli Business Initiative, designed to lure high-tech Israeli firms with security expertise to the borderlands region. Given these interlocking interests, it was no surprise that Arizona's legislators were among the loudest in support of Israel when it launched its genocidal assault on Gaza. They passed a series of staunch pro-Israel resolutions as the genocide unfolded, and in March 2024, as many as one third of the members accepted an invitation to travel to Israel, all expenses paid, while the legislature was still in session.

Nowhere are these shared ties more evident than in the security zone of the borderlands, where DHS agents trained by the Israeli army have jurisdictional powers, and where the high-tech infrastructure is a product, in part, of Israeli industry. In these fragile ecosystems, the military contractors of both nations have also been summoned to combat climate migration. In 2003, a report com-

---

67 Quoted in Matthew Whitaker, *Race Work: The Rise of Civil Rights in the Urban West* (Lincoln: University of Nebraska Press, 2005), 77.

missioned by the Pentagon, titled *An Abrupt Climate Change Scenario and its Implications for United States National Security*, caused a stir by depicting global warming as a threat "worse than terrorism," prompting rich countries to erect "virtual fortresses" around themselves to "hold back unwanted starving immigrants." "Military confrontation," the authors predicted, will be "triggered by a desperate need for natural resources such as energy, food, and water rather than by conflicts over ideology, religion, or national honor.[68] Notably, a remarkably similar forecast of a grim future had been issued almost thirty years before, in a CIA report, *A Study of Climatological Research as it Pertains to Intelligence Problems*, that analyzed the impact of *global cooling*, at that time considered to be the more relevant and potent threat.[69] Like the Pentagon's 2003 report, the CIA's recommended appropriate action in response to its forecast of "the mass movement of people across borders" due to famine and starvation. A few decades later, fortress walls would be built along the US-Mexico border to block these desperate people.

The alarmist rhetoric about catastrophic climate change scenarios was hardly confined to voices from national security circles, always on the lookout to grow their appropriations budgets. In the 1990s, neoliberal intellectuals like Robert Kaplan provided fuel for the border fearmongering. His seminal 1994 essay, "The Coming Anarchy: How Scarcity, Crime, Overpopulation, Tribalism, and Disease Are Destroying the Social Fabric of the Planet," exerted wide influence among policy elites.[70] Shortly thereafter, as part of the

---

68 Doug Randall and Peter Schwartz, Defense Technical Information Center, "An Abrupt Climate Change Scenario and its Implications for United States National Security" (September 2003), https://apps.dtic.mil/sti/citations/ADA469325. Also see Kurt M. Campbell et al., "Age of Consequences: The Foreign Policy and National Security Implications of Global Climate Change," Center for Strategic & International Studies and Center for a New American Security (2007).
69 "A Study of Climatological Research as it Pertains to Intelligence Problems," Central Intelligence Agency (August 1974), https://www.governmentattic.org.
70 Robert Kaplan's influential essay "The Coming Anarchy: How Scarcity, Crime, Overpopulation, Tribalism, and Disease Are Destroying the Social Fabric of the Planet," *The Atlantic*, February 1, 1994, https://www.theatlantic.com.

Prevention Through Deterrence policy of the Clinton administration's Operation Gatekeeper in the mid-1990s, the border wall architecture was "re-designed" to funnel migrants into the "mortal peril" of the deserts and mountains of Southern Arizona.[71]

In the following years, the death toll mounted, prompting a flood of scrutiny and documentation by local humanitarian groups like Humane Borders, Samaritans, and No More Deaths, who left water containers and other supplies along migrant trails in the desert. Even the imminent prospect of dying in hazardous terrains, like Organ Pipe Cactus National Monument preserve, failed to deter border crossers.[72] Consequently, the borderlands saw a massive military build-up, with hundreds of billions of dollars expended on advanced technology (drones, sensors, satellites, radar, surveillance towers, command centers, thermal cameras, aerostat, and facial recognition equipment) accompanied by a suspension of federal laws, regulations, and rights that turned the region into a wartime "zone of exception." The steady growth in corporate contracts fueling this multibillion-dollar market meant that the profit motive was increasingly in the driver's seat.[73]

Among the prime collateral casualties of this militarization are members of the Tohono O'odham, or "desert people", whose lands have seen the highest number of migrant deaths per acre. The drawing of the US-Mexico border after the Gadsden Purchase in 1852 bisected the Tohono O'odham nation, but members in Mexico continued to enjoy access across the binational line until 2001. After 9/11, access to family, schools, doctors, and services got progressively tighter. The nation's lands fell under partial military occupation and territorial control by the CBP.

Like West Bank Palestinians, Tohono O'odham residents on the main Papago Reservation have to submit to military checkpoints to enter and leave the nation's territory, and it is not uncommon for

---

71 Joseph Nevins, *Operation Gatekeeper: The Rise of the Illegal Alien and the Making of the US-Mexico Boundary* (New York: Routledge, 2002).

72 Joseph Nevins, *Dying to Live: A Story of US Immigration in an Age of Global Apartheid* (San Francisco: City Lights/Open Media, 2008).

73 See Todd Miller, "The Many Scandals of the Border Industrial Complex," *The Border Chronicle*, June 20, 2024, https://www.theborderchronicle.com.

darker-skinned members to be racially profiled by CBP agents.[74] Life
on the reservation is subject to high surveillance from ten of Elbit's
surveillance towers that the tribal council agreed to host on Tohono
O'odham land in lieu of a wall. Concern about trafficking on their
lands and drug problems within the community had prompted the
decision, but it has been very divisive, especially among those who
opposed any collusion with federal authorities. Activists don't have
to stretch their imagination to see the kinship with the occupation of
Palestinian lands, extending even to the collaborative relationship of
their leadership with the colonial authorities.

Elbit's experience in administering population control around
Israel's apartheid wall helped the firm win the federal contract to
anchor the electronic, or "virtual," wall that operates a few miles
north of Arizona's physical wall through the construction of fifty-
five integrated surveillance towers. It also prompted the formation
of the Arizona-Palestine Solidarity Alliance, comprising many human
rights groups in the region. Their protests and popular education
efforts have raised awareness about the intimate relationship between
border walls, Indigenous surveillance and displacement, detention,
racial profiling, and militarization in both the US and Israel.[75] On my
visits to Tucson, I met with core activists like Moyheddin Abdulaziz,
a refugee from the village of Emwas [Emmaus], destroyed in the 1967
war, and Sarah Roberts, a retired nurse who is a key organizer with a
number of Tucson groups. They have been successful in mobilizing
campaigns, encouraging Tucson's activists, scholars, and investigative
journalists to undertake deeper analyses of these international ties,
and brokering exchange visits to and from the West Bank.[76] Of all

---

74 Todd Miller, "How Border Patrol Occupied the Tohono O'odham Nation,"
In These Times, June 12, 2019, https://inthesetimes.com.
75 See Anthony Loewenstein, *The Palestine Laboratory: How Israel Exports the
Technology of Occupation Around the World* (London: Verso, 2023).
76 Geoff Boyce et al., "Facing Down Settler Colonialisms: O'odham-Palestin-
ian Struggles," a report by Arizona Palestine Solidarity Alliance (July 2020),
https://www.arizonapalestine.org/; Rebekah Kartal, "Refusing Colonial Forms
of Solidarity on O'odham Lands/the US-Mexico Borderlands," *Antipode* 54, no.
6 (June 2022); Blake Gentry, "Settler Colonialism in Palestine and the O'odham
Homeland," https://www.arizonapalestine.org.

the Palestine solidarity groups in the US, the Tucson-based Arizona alliance is one of the most grounded when it comes to connecting the respective impacts of US and Israeli settler colonialism on land and native peoples. These impacts included water, of course. There are few rivers so alike as the Jordan and the Colorado—each dwindling by the year, with most of their flow diverted to service semi-arid colonies hundreds of miles away.

AT THE WALL

After a 4:30 a.m. wake-up call in downtown Tucson, I joined volunteers with the Tucson Samaritans for the bumpy ninety-minute drive down to the border wall. We were bound for a location just east of the port of entry at Sasabe, whose residents had recently fled after an outbreak of cartel violence. Our team leader, Judy Bourg, a School Sister of Notre Dame with a long record of humanitarian work, warned me that we might not encounter anyone crossing the border. Numbers were down from the October highs, and traffic was unpredictable. In fact, as we drove up and down a seven-mile section of the rollercoaster road by the wall, we came across as many as 150 asylum seekers who had crossed during the nighttime hours. Some were limping, others weeping, or shivering in the morning chill; all were hungry and thirsty. Probably the last thing they expected on setting foot in the US was a warm welcome from the likes of us, nor were they sure what to think, initially, of the food, water, and clothing that we offered. Wary at first, they settled down on receiving firm assurances from the diminutive Bourg that they were legally on American soil, that they had rights as refugees, and that all of us had immigrants in our own family backgrounds. She had even learned local Mayan dialects from her relief work during the Guatemalan dirty wars in the 1980s and was able to namecheck some of their hometowns.

Before the pandemic, asylum seekers were a small number compared to the travelers heading north across the desert, but now the ratio was reversed. All of those we met—from as far away as Albania, Senegal, Mauritania, and Guinea, though most were from Central or South America—were looking to be picked up and processed by CBP. The majority were headed out of state to join family and friends, and they were already rehearsing the harrowing narratives for their asylum cases (though I had no reason to doubt the stories I heard that day about the dangers they were fleeing). Some had been traveling for several weeks, others only a few days, but all of them had made a deal with the smugglers who controlled when and where they crossed. Low-ranking cartel operatives were assigned to cut holes in the fence or help them scale the thirty-foot-high wall. According to *The Border Chronicle* writer Todd Miller, "the cartels have good intelligence, as you would expect from a TCO (transcontinental crime organization), and so the routes they control shift all the time in response to Border Patrol movements." Miller confirmed what volunteers told me, that the cartels routinely buy off CBP agents to ease the passage of drugs across the border.

These organizations diversified their business many years ago, and human trafficking may now have become a bigger business—upwards of $13 billion in annual profits, according to a 2022 estimate.[77] I was told that the fees were around $8,000 or more for Central American asylum-seekers, and even more for migrants who needed additional "services" as they trekked northwards through the desert (Africans will pay much more, and some Asians will pay a VIP price of up to $75,000 for a package trip from their home country). As a result, the cartels were raking in as much, if not more, revenue from the trade in humans than from drugs, smuggled in primarily by Americans returning to the US through official ports of entry.[78] These arrangements deliver many of their clients into indenture; they are so heavily indebted that they are forced into prostitution or drug distribution networks in order to pay off their debts once they are in the US.

Thankfully, the few federal agents we encountered that day were no longer engaged in hunting humans down. Their main purpose now was to pick up and register the asylum-seekers, and they knew that the aid offered by Samaritans made their jobs easier. That said,

---

77 "Smuggling Migrants at the Border Now a Billion-Dollar Business," *The New York Times,* July 25, 2022, https://www.nytimes.com.

78 Jason de León, *Soldiers and Kings: Survival and Hope in the World of Human Smuggling* (New York: Viking, 2024). For an account of coyotes before the cartels took over the business of human trafficking, see Peter Clandestine, *Crossings: Migrants and Coyotes on the Texas-Mexico Border* (Ithaca: Cornell University Press, 2009).

at least one had a blunt MAGA attitude: "What are you here for?" He barked, "What are these people to you?" During their lunchtime shift change, we skirted the law by transporting groups to the main camp where they would be safely picked up, processed by ICE, and taken to Casa Alitas in Tucson, a former Benedictine monastery where volunteers run a shelter on behalf of Catholic Community Services. From there, they would make their way to the airport en route to distant destinations: Illinois, New York, New Jersey, Colorado, California, North Carolina, where they will await their asylum hearings.[79] Some will be deported or sent to the Eloy Detention Center or even further afield; others who don't show up at Casa Alitas will have disappeared into the desert to make their own way north.

Aside from their surreal memories of crossing the border, they will leave behind discarded items of clothing, granola bar wrappers, and plastic bottles, strewn along the twenty-one-mile Arizona section of the thirty-foot-high steel wall erected during the first Trump administration by North Dakota's Fisher Sand and Gravel company (which won its $400 million contract by pitching to the president through Fox News). By the time construction was halted on the day Joe Biden took office, mountains had been leveled, huge scars cut into the landscape, and boneyards of steel pipe littered the desert floor. I passed several gaping holes that were left between the wall's six-foot sections (twenty-three in all) and was told that the Biden administration's policy amounted to little more than "filling the gaps." The futility of this policy was glaringly obvious to anyone who spent more than two hours at the rust-red barrier with holes cut through every night. CBP employed a welder who spent every day patching these holes.

By the time I made a second visit to the same section of the wall in June 2024, Biden had caved to election year pressure and issued an executive order "closing" the border to asylum seekers when the numbers crossing between the ports of entry climbed above 2,500 a day. The order was in clear violation of international refugee agree-

---

79 Elena Mejía and Lisa Beyer, "This Is Where New Migrants Are Going When They Reach the US," *Bloomberg News*, February 28, 2024, https://www.bloomberg.com.

ments, and of the right to seek asylum regardless of how you arrive in the United States, enshrined in section 208 of the Immigration and Nationality Act. Moreover, it was done for political showmanship, since, as with Trump's order, it was unlikely to prevent anyone from crossing. As with every previous crackdown, or federal effort at deterrence, the main consequence was that it forced border-crossers to take more dangerous routes across the desert. Biden's order also came at the worst time of the year, stranding refugees in searing heat, in places along the border without any shade. In 2023, the International Organization for Migration recorded more than 700 migrant deaths along the US-Mexico border (with as many as 8,600 worldwide).[80]

If you have given up your home and sold all your assets to pay for a passage to the border, you will not turn back and you cannot easily return. Those who were en route would keep coming, some unsure about the ramifications of the executive order, and still believing their *coyote* handlers that they would receive a safe haven in the US. On arriving in Tucson in June, I found that the change in the law had already altered CBP treatment of migrants. Casa Alitas, the way station where processed asylees are sent, had seen a decline in numbers from around 800 to as few as seventy or eighty a day. By the end of the month, CBP figures would show that undocumented crossings had fallen to a three-year low.[81] Local news reports suggested that as many as 500 a day, mostly Mexican nationals, were being apprehended along the border and deported in buses to Southern Mexico. While they were supposed to be given a "credible fear" interview on request, reflecting imminent harm or even medical emergency, CBP agents were ignoring this long-standing practice.[82] This amounted to a blunt rejection of the basic right to seek asylum, dating back to the

---

80 "Deadliest Year on Record for Migrants with Nearly 8,600 Deaths in 2023," International Organization on Migration (March 2024), https://www.iom.int.
81 Camilo Montoya-Galvez, "Illegal Crossings at US-Mexico Border Fall to 3-year Low," *CBS News*, July 1, 2024, https://www.cbsnews.com.
82 Emily Bregel, "Border Agents Ignoring Fear Claims, Migrants Say, in Violation of Biden Order Exception," *Arizona Daily Star*, June 15, 2024, https://tucson.com; "Deportations Under Biden's Asylum Ban Reach 500 a Day in Arizona," *12 News*, June 14, 2024, https://www.12news.com.

UN's 1951 Refugee Convention. However, if they were from countries that lacked a repatriation agreement with the US, they could not be returned and therefore stood a better chance of being considered for asylum.

Those best placed to piece together an accurate picture of the new landscape belonged to Tucson's three communities of humanitarians. One of these groups, the Tucson Samaritans, meet in a hallowed location, the Southside Presbyterian Church, where in 1980, the national sanctuary movement was jump-started by the Reverend John Fife (along with Quaker activist Jim Corbett). At that time, the church offered refuge to thousands of Salvadorans and Guatemalans fleeing US-supported death squads in their home countries. Several hundred churches across the country followed suit, building a new kind of Underground Railroad for migrants. In disregard of the ancient custom of sanctuary in places of worship, public authorities prosecuted priests, nuns, and pastors—including Fife—on harboring, smuggling, and conspiracy charges simply for opening their doors. After 9/11, Southside gave another boost to the New Sanctuary movement that sprung up in response to the federal efforts to criminalize migrants. When ICE and CBP were folded into the new Department of Homeland Security, immigration was no longer being seen as a labor issue, it was treated as a national security concern.[83] From the founding energy of New Sanctuary, three key groups were launched in Southern Arizona: Humane Borders (2000), Samaritans (2003), and No More Deaths (2004). They formed a network of volunteers—retirees, medical professionals, youth activists, immigrant advocates, church-goers—who make daily trips into the desert and down to the wall. They welcome new arrivals, save lives through aid, treatment, and clothing, and, most gruesome, they find and document dead bodies (more than 4,000 between 1981 and 2022, according to Humane Borders).[84] Some face criminalization for their

---

83 See A. Naomi Paik, *Bans, Walls, Raids, Sanctuary: Understanding US Immigration for the 21st Century* (Berkeley: University of California Press, 2020).
84 See Jason De León's visual ethnography, *The Land of Open Graves: Living and Dying on the Migrant Trail* (Berkeley: University of California Press, 2015).

good deeds, while others are threatened with physical violence from militia groups.

At the Samaritan meeting I attended in mid-June, members reported on their field shifts at the wall. Water use had increased at the drops, suggesting that there were more "travelers" seeking a route across the desert. The volume of asylum-seekers was fluctuating, though one member estimated that, in Arizona, only 150 asylees were being admitted. Some members thought that the practice of sending across children unaccompanied had already begun to return. The "fair-weather vigilantes" had also returned, emboldened no doubt by Biden's executive order. Recent border-crossers were hailing from new countries such as Sri Lanka, Nepal, Argentina, Liberia, The Gambia, and Belize, and so more foreign language information hand-outs (in Tamil, Hindi, Arabic) were needed. Towards the end of the meeting, a representative from No More Deaths, the sister organization with a younger, hipper, and more diverse membership, presented a plan to coordinate communications between the two groups, along with Humane Borders. It was a significant move, as rifts had opened up in the past between the POC-led No More Deaths and the predominantly white and aging Samaritans (who had been labeled as OWLS—or Old White Liberals—and even less sparingly, as "white saviors"). The urgency of the crisis at the wall had begun to soften, if not entirely override, the divisions.

The next day, I joined Sarah Rogers and Gail Kocourek, one of the longest-serving Samaritans, for our trip to the border wall. Before doing our midday shift, we made a beeline for the Mexican border town of Sasabe to feed dogs who had been abandoned by owners fleeing the cartel violence. Kocourek had opened a community center (Casa Esperanza) in town, but it was vacant now, as the turf war seemed to have become endemic. In the course of our shift at the wall, we restocked the two humanitarian camps along the stretch east of Sasabe, replenished some water drops in the desert and checked on the volume in the fifty-five-gallon drums that Humane Borders operate as fixed water stations. We encountered no migrants, though the Samaritans on the earlier shift had welcomed about twenty-five. Even so, it was unusual to see no one. It was tempting to conclude that Biden's order was taking effect, but a Border Patrol officer whom we quizzed attributed the "lack of alien traffic" to cartel conflict:

"When there's violence on the south side," he shrugged, "things are quiet here."

John Washington, another Tucson-based border writer and author of *The Case for Open Borders*, believed that the smugglers might be in a "wait and see" mode. He said he had seen this pattern before when a new border enforcement was rolled out; the cartels would be very attentive to how CBP officers were interpreting the order. For example, Washington had already heard reports that agents were threatening to separate children from parents. "Biden has sworn that he will never separate children from parents," he pointed out, "but that will be a direct by-product of this new policy, as will many more deaths." When I spoke with him, he had returned from a trip with another humanitarian group Battalion SAR (Search and Rescue), that scours desert trails for lost and missing migrants; they found some human remains. In the summer months that followed, the volume of "encounters" at the border gradually fell, but everyone knew that many more bodies would be found as a direct consequence of Biden's order, and by the much harsher measures adopted by the incoming Trump administration.

On our way back from the end-of-the-wall camp with the Samaritans, two vigilantes on ATVs raced by us. Grizzled and camo-clad, they were out for a joyride, but also to scope out the camp, newly equipped with solar panels and satnav technology courtesy of Starlink. We doubled back to protect the camp and were buzzed by the militia men, intent on wasting our time and patience. Playing army, their behavior seemed juvenile—Kocourek called them "quackadoodles"—but they still meant business. We found out that they had vandalized a temporary shelter en route. Destroying the camps and shooting holes in water containers was routine for these amateur cadets. Aside from the thrill of "owning the libs," their mission is to harass migrants as they make their way to safety, collecting footage that can be sold to Fox News or some other right-wing outlet.[85] In their minds, they are acting out a role of patriotic frontiersmen, defending the nation's

---

85 For a documentary that follows a "citizen patrol" on the Arizona border, see Keegan Hamilton et al., "In Real Life: Crossing the Line," *Scripps News*, December 18, 2023, https://scrippsnews.com.

boundaries. On this theatrical stage, they cannot avoid being cast as armed, unofficial representatives of white supremacy. They appear to be at odds with federal authority over the border, but their MAGA views are shared by many ICE and CBP employees, cementing a longstanding, collaborative relationship between violent paramilitaries and government agents that goes back to the formation of the Texas Rangers and other extralegal groups who have policed non-white populations with impunity.[86] More recently, No More Deaths has documented instances of CBP agents acting as militia proxies by destroying water containers.[87]

Publicly at least, members of the militia groups (like Arizona Border Recon, Mayhem, and Oreo Express) will say that they are needed because the federal government is not doing its job of securing the border. A nation that cannot do so is a failed state. On the other side, the humanitarians are interceding because Washington is not abiding by its obligation to consider and grant asylum claims. When government fails to protect human rights and imperiled lives, civil society must step up.

Leaving aside the moral urgency of this intervention, some advocates like to point out that immigrants, and especially the second generation, provide a net economic boost to the countries that absorb them. A 2023 study showed that, in 2021, they generated $524.7 billion in total taxes, including $346.3 billion in federal taxes, and $178.4 billion in state and local taxes, and held $1.4 trillion in spending power—a huge boost to GDP.[88] Immigrants contribute much more

---

86 Kelly Lytle Hernández, *Migra! A History of the US Border Patrol* (Berkeley: University of California Press, 2010); Monica Muñoz Martinez, *The Injustice Never Leaves You: Anti-Mexican Violence in Texas* (Cambridge: Harvard University Press, 2018); and Greg Grandin, *The End of the Myth: From the Frontier to the Border Wall in the Mind of America* (New York: Metropolitan Books, 2019).

87 No More Deaths, "Interference with Humanitarian Aid," is the second part of a serial analysis of how the practices of CPB, ICE, and local police have resulted in deaths and disappearances. *Disappeared: How the US Border Enforcement Agencies are Fueling a Missing Persons Crisis*, www.thedisappearedreport.org.

88 American Immigration Council, "American Immigration Council Maps the Impact of Immigrants with New Data on the United States and All 50 States" (November 13, 2023), https://www.americanimmigrationcouncil.org.

in payroll tax than they receive in social security benefits, they create more jobs than they "take," and their capacity to revitalize economically depressed communities is well documented. Those who beat all the odds to make it across the border tend to have skills, education and social capital, and lower rates of ill-health and obesity than most of the US working-class population, and so they are much more likely to be an asset than a burden. The flow of remittances they send to their countries of origin is now a vital source of revenue and a key component of the global economy. Contrary to the fear-mongering conservative talking point, crime is generally lower in immigrant communities. Where crime *is* prevalent, however, is along the border wall itself because of the cutthroat business in human smuggling and other forms of trafficking.

The problem with such arguments is that they reinforce the image of immigrants as the most productive, law-abiding, and healthy laborers of all. This "good immigrant" narrative is interdependent on the mythology of the "bad," shiftless immigrant, and it is implicitly racist in its implications for native-born workers who do not measure up by comparison, and who end up as a welfare "burden" or as wards of the carceral state. As Ashley Dawson has argued, calling attention to the benefits accruing to "receiving countries" glosses over their own reparative responsibilities to donor countries, in particular, their obligations to repay climate debts in the form of adaptation funding.[89]

Until the late 1990s, the US-Mexico border was almost devoid of infrastructure, and there were only a handful of walls across the world. Now there are several dozen, cordoning off countries that share a border with less affluent populations, including Austria, China, Hungary, Iran, Israel, India, Morocco, Poland, South Africa, Serbia, and Turkey. To stop the migrant boats from getting across the Mediterranean Sea (now a mass maritime graveyard), the EU has "exteriorized" its borders by providing a suite of loans and grants, through the Emergency Trust Fund and others, to Turkey, Egypt, Tunisia, and other "transit states" in Africa on condition that they seal their borders. Indeed, development aid to poor countries is

---

89 Ashley Dawson's chapter on "climate apartheid" in *Extreme Cities: The Peril and Promise of Urban Life in the Age of Climate Change* (London: Verso Books, 2017), especially pages 228–31.

increasingly being tied to their willingness to stem the flow of refugees.[90] These agreements are evocative of the decades when Western development aid was tied to birth control in poor countries. Such conditions are no longer deemed acceptable, but the policy of bribing countries to stop migrants is cut from the same cloth. Under these arrangements, the well-being of those on the move are often sacrificed by strongmen in power who have designs on the aid. Libya's Muammar Gaddafi, who was the first to sign such an agreement with Italy's Silvio Berlusconi in 2008, threatened to "turn Europe Black" if more funds were not forthcoming. The outcome was a network of gruesome detention centers, where detainees can be held indefinitely and are subject to rape, torture, and labor trafficking.[91]

Far from being a guarantor of security, the even vaster sums expended by the US and European nations (and Australia) on border and immigration enforcement are more likely to generate extreme harm from violations of human rights—deaths from heat, thirst, starvation or drowning, torture, rape, indenture, indefinite incarceration, and discriminatory violence from native-born communities. To add insult to injury, the harms are visited mostly on people who are moving from situations of extreme peril in order to be safe.

The case for open borders has been made on the libertarian right as well as the left, on moral and economic grounds alike.[92] If this abo-

---

90 See "How Europe's Fear of Migrants Came to Dominate its Foreign Policy," *The Economist*, March 27, 2024, https://www.economist.com.
91 Ian Urbina, "The Invisible Wall: Inside the Secretive Libyan Prisons That Keep Migrants Out," *The New Yorker*, December 6, 2021, https://www. newyorker.com. Lesser, but significant, forms of exploitation have occurred in privately owned and run ICE detention centers within the US, in violation of the prohibition on forced labor in the Trafficking Victims Protection Act. Victoria Law, "End Forced Labor in Immigrant Detention," *The New York Times*, January 29, 2019, https://www.nytimes.com.
92 The libertarian or conservative case has been made by Bryan Caplan, in *Open Borders: The Science and Ethics of Immigration* (New York: First Second, 2019) and Jason L. Riley, in *Let Them In: The Case for Open Borders* (New York: Gotham, 2008), while John Washington argues from the left in *The Case for Open Borders* (Chicago: Haymarket, 2024), and Harsha Walia makes the abolitionist case in *Border and Rule: Global Migration, Capitalism, and the Rise of Racist Nationalism* (Chicago: Haymarket, 2021).

litionist scenario ever comes to pass, it will be no more than a return to the status quo ante, when human mobility across large distances was a normal feature of planetary life. Indeed, population geneticists have demonstrated that migration, and not sedentary settlement, is the dominant driver of human evolution. Yet, in recent decades, the "specter" of people on the move is everywhere tapping into deep reserves of hostility.[93] The mass of humanity who travel internationally as tourists do not generate instability, disease, or disorder, and yet the prospect of poor people crossing the borders of a wealthier nation is viewed as a calamitous breach of its sanctity. Nowhere are these pathologies more explicit than on the southern border of the US.

The frontline work done by Samaritans, No More Deaths, and the other humanitarian groups is driven by a degree of compassion that has been ruthlessly purged from the US border system. Stopping people from dying is the bare minimum when it comes to humane responses. But this kind of work is also a form of mutual aid and collective care. It draws on the belief that access to land and territory is a common right to be shared and not a property claim on the part of those who happen to be born on the right side of a national border. In the current configuration, public power, manifested in the form of the $30 billion budget of ICE and CPB, is on the wrong side of history, and very much at odds with this notion of a caring commons. Nor is it the preserve of the hard right. Bipartisan politics built the border regime, with a large share of the responsibility lying with Democrats like Bill Clinton, who initiated mass deportations as a result of his 1996 Immigration Reform and Immigrant Responsibility Act, or Obama, who earned the moniker of "deporter-in-chief," or Biden, who shredded the principle of asylum with his executive order. Over the last two decades, every administration, whether red or blue, has poured more money into the physical and personnel infrastructure for blocking, mistreating, and expelling migrants. Despite Trump's hyperbolic campaign bluster about mass deportations, some border hands, like Todd Miller, thought that his second administration could

---

93 Science journalist Sonia Shah offers an instructive view on the history of migration in *The Next Great Migration: The Beauty and Terror of Life on the Move* (New York: Bloomsbury, 2020).

hardly do worse than Biden, who deported more than Trump had, and who left office "as the king of border contracts," having enriched countless private companies through the issuance of "21,713 border enforcement contracts, worth $32.3 billion, far more than any previous president."[94]

Trump's first weeks in office saw a flood of anti-immigration executive orders: declaring a national emergency at the border, halting all refugee admissions, ending humanitarian parole for immigrants from "favored status" countries, abolishing birthright citizenship, limiting Temporary Protected Status, denying public benefits to unauthorized immigrants, targeting sanctuary jurisdictions, and demanding local cooperation for immigration enforcement. A flurry of theatrical deportation raids followed, along with a host of more costly measures, including the expansion of detention facilities, the deployment of an additional 1,500 troops to the borderlands, increased recruitment of ICE and CBP officers, and land acquisition for building even more border barriers. On my third visit, in March 2025, the border zone was almost devoid of activity. No asylum-seekers were visible, though the frequent use of water stations along migrant trails suggested that the passage of "travelers" trying their luck at crossing the desert had picked up.

The month before, the Senate approved an additional $175 billion for border security, including funding for housing migrants at the notorious military instillation on Guantanamo Bay in Cuba, which a 2019 *New York Times* analysis found, despite its wretched condition, to be the world's most expensive prison, costing roughly $13 million per detainee each year.[95] Within a few weeks of my visit, the Trump administration requested another $45 billion for new detention facilities, transportation, security guards, and other adminstrative services.

Imagine these vast public resources being used instead to welcome people fleeing danger, poverty, and environmental ruin, in the

---

94 Todd Miller, "The Mass Deportation Handoff," *The Border Chronicle,* February 6, 2025, https://www.theborderchronicle.com.
95 Carol Rosenberg, "The Cost of Running Guantánamo Bay: $13 Million Per Prisoner," *The New York Times,* September 16, 2019, https://www.nytimes.com.

spirit of Emma Lazarus' remarkable lines engraved on the Statue of Liberty: "Give me your tired, your poor, your huddled masses yearning to breathe free." Consider how these vast sums could be spent on helping new arrivals settle and integrate into a society that has fallen below its population replacement rate, and whose economy has come to rely on their "essential labor." The US has always used its public power to facilitate (white ethnic) immigration, even as it has been deployed to decimate Indigenous populations. The granting of vast areas of appropriated Indian land to emigrants in the nineteenth century is the formative example of this two-sided policy, while the Chinese Exclusion Act, passed not long after, initiated an enduring tradition of racial bars. The more recent history includes the resettlement of Cambodians and Vietnamese after the war in Indochina in recognition of federal responsibility to refugees from American overseas wars. Other favored groups have included Afghans, Cubans, Syrians, Venezuelans, and Ukrainians—all of them from countries embroiled in US imperialist conflict. Refugees from these countries were shown hospitality, but those on the other side of the US war machine are not welcome.

No one who has anything useful to say about the US-Mexico border will conclude that the challenges can be met by the American polity acting alone. Any humane solution, as the Sierra Club's Erick Meza put it, must be "a transboundary one, requiring binational work with Mexico at the very least." But ultimately, he urged, "what has been happening in these countries needs a proper prevention strategy—because the US interventions in these countries have devastated the livelihoods of so many people and forced them to move." This kind of transboundary cooperation must extend far beyond the policy of "asylum cooperative agreements" whereby rich countries in temperate latitudes pay poor ones to stop people from traveling north. As Washington put it, "immigration routes are much too complex for any one country to deal with."

What other steps need to be taken? High CO_2 emitters should meet their obligations to populations least responsible for the climate emergency by settling historic carbon debts. As climate migration increases, a new international legal regime should recognize the right to migrate and the right to refuge as well as the right to remain. And we need to build a stronger culture of mutual aid to combat the mentality of proprietary hoarding which requires militarized national borders to be wielded as instruments of violence and extinction.

Achieving these goals will require an effort to link struggles that are often seen as separate. The Sanctuary movement, which bursts into life intermittently, and carries the memory of its roots in anti-imperialist critique, saw another revival under Trump 2.0. But, as Silky Shah reminds us, border politics are also intertwined with US mass incarceration. Immigrant detention centers, where detainees can linger from two to five years, are not so far removed from the carceral archipelago of state and federal prisons and are often operated by the same private firms profiting royally from the new deportation drive.[96] So, too, the unlawful detention of migrants in military prisons like Guantanamo underscores the continuity with martial discipline and expansion.[97] Prison abolition necessarily overlaps with border abolition. A variety of activist groupings—including the Dreamers, Black Lives Matter, and Critical Resistance—have learned from each other that policing, whether in the city streets, at the border, behind the walls, or in countries within the orbit of US empire, is the common enemy.

---

96 Silky Shah, *Unbuild Walls: Why Immigrant Justice Needs Abolition* (Chicago: Haymarket, 2024).

97 See Jenna Loyd and Alison Mountz, *Boats, Borders, and Bases: Race, the Cold War, and the Rise of Migration Detention in the United States* (Berkeley: University of California Press, 2018).

Patriotic monument in Jinggangshan City, Jiangxi province, April 2024.

4

CHINA

THE LAST GENERATION?

When I lived in Shanghai in 2002, the city was still in the throes of China's national one-child policy. For urbanites, and especially those employed by the state, there were few exceptions allowed, although those with means could always pay a fine for having more than one offspring. Parks and playgrounds were busy with parents and grand-parents shepherding around their "little emperors" [*xiao huangdi*], the Chinese term for a pampered single child. The policy was looser in the countryside, where couples could "try" for a boy if their first-born child was a girl. Many of the "illegal seconds," if they were female, were still being abandoned, and the country's orphanages were overflowing. In order to address the yawning male-to-female imbalance, portable ultrasound machines, which had been widely used to select for gender, were banned that same year. By the time of my return in 2024, the government's interest in family planning had turned completely upside down. China's overall population had shrunk for two years running, and the fertility rate was among the lowest in the world. For the past ten years, the Communist Party of China (CPC) had been running a multipronged campaign to boost births. As in other countries facing a dismal "demographic winter," their efforts were failing. The Chinese General Social Survey of 2021 found that almost half of the women born after 1995 said they desired one or no children.[1] According to a more recent survey, 18.69 percent of women in the age range of twenty to thirty-nine in Shanghai expressed the intention of having no children.[2] Some wisecrackers had begun to refer to these women as the "last generation."

During the month I spent in the city, I conducted lengthy interviews with fifteen college-educated women who fell within that age group. Almost all grew up in single-child families. Most had no children of their own, while a few had one child. In the course of our conversations, they talked about their upbringing, the economic and social pressures in their lives to start a family, and the decisions they were making as a result. They were all quite passionate on the topic, and had strong feelings about the defensive position they found themselves in. Unlike their grandparents and parents, who had been asked to sacrifice (or "eat bitterness") for the national good, none of them felt responsive to the government's attempts to encourage them to reproduce. Quite the opposite, in fact. They all spoke of the loss of public trust in the authorities, especially Shanghai's officials, after the harsh policing of the general population during the long zero-COVID lockdowns. The days when the CPC could mobilize the people for a nation-level campaign seemed to be over. Nor was Xi Jinping's increasingly autocratic rule helping matters.[3]

Two of the childless women I interviewed described their difficulties finding an appropriate husband and said that adoption might be an option for them. Fang Gao, a thirty-two-year-old, third-

---

1   Cited in "China's Low-Fertility Trap," *The Economist*, March 21, 2024, https://www.economist.com.

2   Haobo Deng and Yuxuan Jia, "Over 5% Chinese Women Embrace Life Without Children, Surging in the 2010–2020 Period," *East is Read*, March 25, 2024, https://www.eastisread.com.

3   One interviewee expressed her scorn when, the day before our meeting, she received an official text message (commemorating National Security Education Day) warning about foreign spies pretending to be academics.

generation Shanghainese, imagined that she might someday meet a child with whom she felt a strong connection, even though infants in need of parents were in short supply these days. This connection, she said, would be a manifestation of *yuanfen*, a folk belief, widespread in China, that fate or destiny can bring two people together. "I don't know where or when that could happen," she mused, "but I could be influenced by it enough to adopt the child." Yuanfen is often invoked as an explanation for a variety of serendipitous events or relationships, but it has a particular significance for the presumed bond between an orphan and an adoptive parent. This quasi-mystical connection was encouraged during the heyday of international adoption from China in the form of an "invisible red thread" that is spun at birth and connects a parent with the child as if by destiny. The conviction that a child was meant to be with a particular parent was much criticized at the time. The corollary was that the birth parents were destined to "lose" the child, when, in fact, the circumstances of abandonment were almost entirely the result of socioeconomic and political conditions, and a source of great pain and grief for them.[4]

CRADLE OF THE REVOLUTION

For domestic adoptees, there are less restrictions today than there were during the one-child era, when it was much easier for foreigners to adopt.[5] I was one of the latter. During the year we spent in Shanghai with our biological daughter, my partner and I heard many of

---

4 Kay Ann Johnson interviewed many of the parents who were forced to give up their babies in *China's Hidden Children: Abandonment, Adoption, and the Human Costs of the One-Child Policy* (Chicago: University of Chicago, 2016). See also Mei Fong, *One Child: The Story of China's Most Radical Experiment* (Boston: Mariner Books, 2016), 189.

5 From 1992 to 2024, when the programs were ended, as many as 160,000 babies were adopted out from China. Farah Master, "China Stops Foreign Adoptions of its Children After Three Decades," *Reuters*, September 6, 2024, https://www. reuters.com. See also Mari Manninen, *Secrets and Siblings: The Vanished Lives of China's One Child Policy* (London: Zed Books, 2019). For a related memoir of an adopter, see Karin Evans, *The Lost Daughters of China: Abandoned Girls, Their Journey to America, and the Search for a Missing Past* (New York: Jeremy P. Tarcher/ Putnam, 2000).

the harrowing stories about the gruesome circumstances of orphans. Some Shanghainese acquaintances were aware of the revelations in two BBC documentaries that had exposed the callous neglect of babies in China's orphanages. The first film, *The Dying Rooms* (1995), focused on provincial institutions, while the sequel, *Return to the Dying Rooms* (1996), investigated abuses documented in a Human Rights Watch (HRW) report about the Shanghai Social Welfare Institute, in the country's most culturally advanced and cosmopolitan city. Former employees and inmates testified about the blatant mistreatment on the part of senior staff, including their policy of "deliberate starvation" of select infants in order to control the orphanage's intake and to make the mortality pattern look like natural deaths. The HRW report concluded that up to four out of five orphans died within their first year, and that the death rate was above 50 percent in thousands of such facilities across the country.[6]

As many commentators pointed out, it was impossible to verify these estimates, and the international fallout from the report and the films precipitated an information blackout from the Chinese government that prevented any further investigations.[7] But the disclosures, and subsequent ones alleging child-trafficking, helped to reinforce the prevailing Western perception of China's one-child policy as a pitiless, barbaric enterprise.[8] The campaigns on the part of Western population control advocates had resulted in harsh top-down measures in other poor Asian countries in the 1960s and 1970s. But the stories coming out of China from the 1980s were amplified by anti-

---

6   Human Rights Watch, "Death by Default: A Policy of Fatal Neglect in China's State Orphanages" (January 1996), https://www.hrw.org.

7   Human Rights Watch responded to the criticism in "Chinese Orphanages: A Follow-Up" (March 1996), https://www.hrw.org.

8   In 2005, allegations surfaced of trafficking in Hunan, followed by stories in other provinces. Brian H. Stuy argues that up to half of the orphanages involved in international adoption programs (where adopters were required to make hefty donations to the institutions and pay fees to officials) were buying babies. "Open Secret: Cash and Coercion in China's International Adoption Program," *Cumberland Law Review* 44, no. 3 (October 2014). The allegations were recirculated in the controversial 2019 documentary *One Child Nation*, directed by Nanfu Wang and Jialing Zhang, that rekindled Western prejudices.

communist sentiment and also helped to jumpstart the anti-abortion movement in the US.

By the time we lived in Shanghai, we had already decided that if we had another child, it would be through adoption. We studied the challenges involved in transracial adoptions in general and learned, sadly, that many Western adopters were encouraged to believe that they were "rescuing" infants that Chinese parents had been forced, heartlessly, to abandon. But the silence we encountered that year about the country's orphanages (estimated to number as many as 40,000 at the time) combined with the undeniable need of the babies themselves for caring parents, convinced us to adopt in China.

Three years later, we traveled to Jiangxi province to pick up Stella Kang, who had been named by her orphanage, in Taihe County, as Dongfang Kangyu. As was customary, we traveled with a group of adopters, many of whom were American Christians whose churches had spurred them on to "save" children and their souls. For some of them, the "invisible red thread" narrative had a religious overlay; God had chosen their child for them. The opaque circumstances of abandonment—most babies were left outside the orphanage gates—also meant that it was not uncommon for adopters to harbor fantasies about the social origin of the babies they would take home with them. More than one parent in our party expressed a belief that their daughter was from a genteel family that had suffered hard times during the Maoist era when bourgeois households lost many of their privileges. Their hope was that the child's social grace and sense of class would be restored once they were reared in the US.

We had a different kind of fantasy about Stella's family. Taihe County is not far from Jinggangshan, the mountainous region officially known as the "cradle of the Chinese Revolution." This is where the Chinese Red Army (predecessor of the People's Liberation Army) was first formed, and where Communist forces established the country's first self-governing "soviet" in the late 1920s. After Chiang Kai-shek's Nationalist Party (Kuomintang, or KMT) betrayed and massacred their Communist allies in Shanghai in 1926, Mao Zedong, Zhu De, Lin Biao, Zhou Enlai, Chen Yi and other party leaders retreated to the countryside, setting up their first revolutionary base in mountainous areas that were easier to defend. The Chinese Soviet Republic was proclaimed in Ruijin in November 1931, and its rule was extended beyond southern Jiangxi to western Fujian and parts of Guangdong province for almost three years. Hemmed in finally by the encirclement campaign of the KMT forces, the remnants of the communist divisions set out on the Long March to Yan'an in 1934.

Perhaps, we told ourselves, our daughter's family had belonged to the local peasantry who were radicalized by the experience of this socialist alternative (some historians have even argued that the local peasantry initiated the revolutionary movement before Mao and others arrived).[9] If so, they would have been left behind when the core communists departed, and would likely have lived through the Confucian neofascist experiments of the KMT leaders, who chose Jiangxi as a model for their New Life Movement in the latter half of the 1930s.[10] Subsequently, they would have participated in the massive social upheavals of the next several decades—the great Maoist experiment of agricultural collectivization, the ill-guided Great Leap Forward and the famine it engendered, the tumultuous Cultural Revolution, the advent of the socialist market economy under Deng Xiaoping, and of course, the one-child policy responsible for Stella's adoption experience.

Initiated in 1990, China's international adoption program peaked the same year we adopted Stella and it declined sharply thereafter, as the criteria for foreign adoption became much stricter. After the two-child policy was introduced in 2014, the supply of abandoned babies began to dry up, and many orphanages were shuttered or repurposed as elderly care facilities. China terminated its international adoption program in 2024.

Stella was now eighteen years old and she wanted to visit. So, in late April 2023, she and I set off for Jiangxi, which had been one

---

9   Joseph Fewsmith argues that the local Jinggangshan peasantry made the first revolutionary moves, by destroying clan leadership structures. Their bottom-up efforts at the Donggu revolutionary base were subsequently subsumed into the more hierarchical, Leninist discipline of the incoming Communist party leadership. *Forging Leninism in China: Mao and the Remaking of the Chinese Communist Party, 1927–1934* (Cambridge: Cambridge University Press, 2022).

10 Arif Dirlik, "The Ideological Foundations of the New Life Movement: A Study in Counterrevolution," *The Journal of Asian Studies* 34, no. 4 (August 1975): 945–80.

of the provinces where the birth control policy was most strictly enforced and had therefore hosted a great number of orphanages. In the mid-2000s, it was the province with the most gender disparity, with as many as 143 males to 100 females,[11] and in 2014, it was one of the first, along with Anhui and Zhejiang, to be chosen to pioneer the new two-child policy.[12] Four years later, Jiangxi generated a new alarm, when the local government initiated new guidelines for when women could get abortions; after fourteen weeks, they needed three signatures from medical personnel before having the procedure.[13] It was the first sign that the state's family planning apparatus would continue its vigilance, this time in the service of *encouraging* rather than restraining births. Officials insisted this measure was intended to deter couples from aborting a female fetus in hopes of having a boy, but they acknowledged that government pressure to keep the birthrate up was also a factor.[14]

One other local nugget of knowledge we took with us was about the high incidence in Jiangxi of "little-daughter-in-law" adoptions made by couples who were intent on an arranged marriage to an existing biological son. Often undertaken to avoid steep "bride prices," this kind of adoption was most prevalent in Jiangxi and Fujian. According to one 2015 study, these adoptions accounted for at least 6 percent and as many as 10 percent of girls in 1994, its peak year.[15] This practice was an old, though somewhat disparaged, tradition known as "minor marriage," whereby the adopted female infant could be groomed for obedience in marriage while also performing

---

11 Wei Xing Zhu, Li Lun, and Therese Hesketh, "China's Excess Males, Sex Selective Abortion, and One Child Policy: Analysis of Data from 2005 National Intercensus Survey," *British Medical Journal* 338 (April 2009): b1211.

12 Xu Wei, "Zhejiang, Jiangxi and Anhui Pioneer in Two-Child Policy," *China Daily*, February 10, 2014, https://global.chinadaily.com.cn.

13 Lily Kuo, "China: New Rules to Prevent Sex-Selective Abortions Raise Fears," *The Guardian*, June 22, 2018, https://www.theguardian.com.

14 Steven Lee Myers and Olivia Mitchell Ryan, "Burying 'One Child' Limits, China Pushes Women to Have More Babies," *The New York Times*, August 11, 2018, https://www.nytimes.com.

15 Yuyu Chen, Avraham Ebenstein, Lena Edlund, and Hongbin Li, "Girl Adoption in China—A Less-Known Side of Son Preference," *Population Studies* 69, no. 2 (2015): 161–78.

domestic labor for the household. Its recurrence at the height of the one-child policy showed how longstanding regional customs offered an adaptable response to the widespread rise in abandonment of girls from the 1980s onwards.

Whether or not the tradition lives on, the monetary value of a bride price had continued to rise, and Jiangxi was at the top of the nationwide list. The cost of this reverse dowry averages more than $53,000 and can go as high as $116,000, not including the house and car which are increasingly part of the package.[16] The reason most cited for the exorbitant price is a shortage of brides stemming from Jiangxi's gender imbalance, while in rural areas, the market has been overheated due to the exodus of women seeking better job opportunities in cities and the disadvantages faced by farmers' sons in the matchmaking stakes. Needless to say, government officials, tasked with boosting fertility, were trying to regulate the cost of marriage by putting a cap on the bride price.

Though we had no reason to expect any welcome at all, Stella was warmly received at her orphanage, the Taihe Social Welfare Institute. A streaming neon display at the entrance announced her return. The director, Lihua Zhou, had postponed an international flight to visit her daughter in Germany to be present, and a photographer was there to document every detail of our day with the staff. "It is not often that the children return here," Zhou explained, "and there have been none since before the pandemic, so we are all very interested in her." Indeed, there was a palpable excitement on the faces of staff, tinted, for some of them, by nostalgia for the heyday of the institution. Stella's file, it turned out, was in impeccable order. It recorded all of her health particulars during the first year of her life that she spent there and contained photographs we had not seen before. One of the older staff members present recalled that Stella had been found outside the orphanage by her own father-in-law. Hearing about the visit, one of Stella's caregivers rushed from her bank workplace across town to join us.

---

16 China Desk, Lianhe ZaoBao, "China is Cracking Down on Exorbitant Bride Price Rates to Save Marriages," *Think China*, February 28, 2023, https://www.thinkchina.sg.

The wing of the institution once allocated for infants still hosted a few teenagers with special needs, who were consumed with interest in us. But the rest of the residents were elders, and they seemed entirely indifferent to us and to the purpose of our visit. They are the face of China's graying future: across the country, institutions (including kindergartens) that overflowed with infants fifteen years before are now filling up with seniors. Twenty-one percent of the country's population (or 297 million) are now over sixty, and by 2040, an estimated one third will be. Footing the bill for their care is a paramount concern, especially in rural areas like Taihe, where pension benefits are thin because of the spread of informal employment. Nor are the elderly assured traditional forms of support from offspring who are no longer so numerous or devoted in their sense of filial duty.[17]

The director, who had been in her position since the late 1990s, told us there were three categories of residents at the institute. Those who had served as military cadre were non-fee paying. So, too, were the seniors with no retirement income and no children to look after them. The others, who had children and some income, had to pay the fees. But while other welfare institutes had accepted private management, the Taihe Social Welfare Institute was still a wholly public agency. It had been a model unit, selected for international adoptions. Our taxi driver told us that he had often ferried babies found in other counties to the orphanage, presumably in hopes they would be treated well and adopted. On learning this, Stella realized that, even if she wanted to, the challenge of tracking her birth parents would be much harder than she had bargained for.

The director had ordered a banquet lunch to be prepared for us. The offerings included Taihe's signature delicacy—the black-boned Silkie chicken, reared exclusively in these parts for more than two thousand years, and a well-known, if rare, dish in the repertoire of Chinese cuisine. After feasting, we were escorted to the county's memorial museum to the martyrs of the 1930s military campaigns.

17 See Vanessa Fong, *Only Hope: Coming of Age under China's One-Child Policy* (Stanford: Stanford University Press, 2004).

The day before we had visited the revolutionary base in the Jinggang-shan mountains. A site of patriotic education, it is a mainstay of "Red Tourism," and receives busloads of domestic visitors, though few foreigners. Heroic sculptures of military figures are carved out of the cliffs en route, and the carefully landscaped destination sites are festooned with symbols and images of well-known leaders, from the Jiangxi Soviet's era (1931–34) of CPC revolutionary history.

By contrast, the Taihe museum is focused on local history and personages. Unlike the famous faces of Jinggangshan who are still honored today by cadres and nationalists, the martyrs who are metic-ulously commemorated in the Taihe museum were small-time par-ticipants in the struggle with the KMT forces. Many gave their lives before they had families, and so there may have been few direct descendants of these men and women who became organic com-munists at a time when the cause was still fresh and the odds against them prevailing were overwhelming. During the Second World War, Taihe had another moment in the sun, when Jiangxi's govern-ment relocated there from Nanchang to seek refuge from Japanese bombs. Today, it boasts towering high-rises, like any other residential center in China, but just outside of town lies its fertile agricultural mainstay—mile after mile of terraced paddy fields, yielding two rice harvests a year.

Several of the institute staff had a child, but no one we asked had more than one. The government's pro-natalist campaign didn't seem to be working here either. Jiangxi's municipalities were not promising the same inducements on offer in cities like Shanghai—prolonged parental leave, cash payments for a second or third child, or subsidized housing. In provincial towns like this, where red banner slogans announcing directives about birth planning were once ubiq-uitous, "population engineering" was now done through messaging about having a "correct attitude toward parenting." Or it came from parents and grandparents, ever desirous of progeny. As it happens, on my return to Shanghai, I interviewed a woman whose best friend was from a rural Jiangxi family. She had recently "escaped" to California and cut ties with her family, because she could not abide the relentless cajoling from her parents to produce a child. Her brother was the center of their universe, and she had been brought up to prioritize, and contribute to, his well-being above all else. I was told that the "family's patriarchal values" had "poisoned" her thoughts about having a family of her own. A lifetime of deference to him, and to men like him, had turned her off.

MORE IS LESS

Shanghai's "apron husbands" are supposed to be different; they have long been sought after as obedient and domesticated mates. The city's enduring reputation for men who do their share of household labor is still summed up in the Shanghainese slang phrase "*ma da dao*" [shop, wash, cook]. If only they could give birth on their own they would be the new "model workers" for CPC officials who still select moral and industrious exemplars for everyone else to emulate. China's notable gender imbalance means that a sizable number of men are now "bare branches," unable to find wives and help to stem the fall in birth rates that has engendered a state-level crisis. To make matters worse, their semen quality has deteriorated. Whether or not a result of pollution, as has been suspected, one study found that only a third of samples donated to a Shanghainese sperm bank met the WHO's fertility standards.[18]

Women, however, are the ones feeling all the pressure from the government. For college-educated, single women like Zhengyun Li, the challenge of having a child with a suitable husband—"I really don't think it's possible for me to find one in Shanghai"—is not simply a personal concern. As China's president, Xi Jinping, himself has put it, " The birth of a baby is not only a matter of the family itself, but also a state affair."[19] Li is well aware that "having a child, with someone like me as a parent, would be very good for this economy," but explained

---

18 Stephen Chen, "Smog Crisis in China Leads to Increased Research into Effect of Pollution on Fertility," *South China Morning Post*, December 11, 2013, https://www.scmp.com; Rachel Montgomery, "Shanghai Smog to Blame for Subpar Sperm, Says Clinic Head," *Bionews*, November 11, 2013, https://www.progress.org.uk. It is likely, however, that the impact of pollution on fetal health was greater, resulting in a higher rate of miscarriages. See Timothy Beardson, *Ageing Giant: China's Looming Population Collapse* (Oxford: Signal, 2021), 16.

19 Steven Lee Myers and Olivia Mitchell Ryan, "Burying 'One Child' Limits," *The New York Times*, August 11, 2018, https://www.nytimes.com.

that "I would need the right kind of husband for that." Under the government's efforts to grow households, single women like her are no longer officially prohibited from having children, but they balk at the soaring costs of parenting, especially the largely privatized care services for children under the age of three. "Even with my well-paid salary, I would struggle to pay the bills," she explained, "and if the government wants me to have children, which I know is their policy, they should be helping me much more than they are prepared to do."

Li, who works as a school administrator, acknowledges that she has high standards for a potential mate—"Why shouldn't I?" she laughs—and that she would consider adopting a child before raising a biological one with a man who falls short. "I have options," she points out, "that few women in China have ever had." Like my other interviewees, her sense of independence was, in part, a consequence of the one-child policy, which had harsh consequences for many rural women, but gave rise to a well-educated, cosmopolitan generation of urban women. This cohort stepped up into valued professions and its members were driving a significant consumer boom—labeled the "sheconomy"[20]—but their reluctance to perform the child-bearing role of Confucian mothers was a drag on the state's future economic growth, including the guarantee of sufficient tax revenue for the purpose of "regime survival." Indeed, the numbers who have chosen to remain single and childless are rapidly increasing. By 2022, 56 percent of the nation's women between the age of twenty-five and thirty were single, and 20 percent between the age of thirty and forty.

While the most recent declines in births are noteworthy, the country's fertility rate has been steadily falling since the peak of 6 percent in the mid-1960s. In the mid-1990s, it dropped below the replacement rate of 2.1, and reached 1.16 in 2023, the world's lowest with the exception of Singapore, Taiwan, and South Korea which has the lowest rate at 0.72. The downward turn in the country's overall population, which began in 2022, was the first decrease since 1961, after the famine that followed Mao's Great Leap Forward. Not surprisingly, there is official concern about the long-term availability of

---

20 Zhang Nan, "What's Fueling the Rise of 'Sheconomy'?," *Global Times*, June 2, 2024, https://www.globaltimes.cn.

the abundant labor supply that fueled China's economic rise. Many other nations, especially in Asia, are in the same boat, but China is still the locomotive of the global capitalist economy. Despite losing some of its manufacturing to lower-wage countries in Southeast Asia, it still produces a third of the goods consumed globally and is the largest trading partner for more than 120 countries. The recent and ongoing failure of officials to persuade women to produce more children for the labor force is a source of intense anxiety, not just for the managers of the "workshop of the world," but also for its numerous foreign investors.

It is often assumed that the declining birth rate is a direct consequence of the one-child policy, and that smaller families simply became normative as a result of the state's stringent birth control efforts. Yet just as the actual impact of that policy on China's population is still a matter of debate—the birth rate was falling anyway—so too, other East Asian countries with no population control policies have shown almost the same demographic pattern. It is also tempting to interpret the reluctance to procreate as a motivated response to the overreach of state authority, or, as one commentator has put it, as a form of "internalized civil disobedience."[21] In the 2010s, for example, government media organs started a whispering campaign against "leftover women" who had aged out of the traditional childbearing window. This effort at overt shaming was met with widespread popular derision, especially in urban quarters.[22] It did not help that the campaign of stigmatization was colored by empathy for their male counterparts—the leftover men, or "bare branches." Even so, there is little evidence that Chinese women are opting out of motherhood *in defiance* of the overtures of government officials.

As I learned from my interviews, college-educated urban women are the least willing to marry down, hence the challenge of finding a match. Their careers and independent incomes afford them freedoms outside of the orbit of patriarchal norms, and so they are under less social and economic pressure to start a family. The government has

---

21 Evan Osnos, "China's Age of Malaise," *New Yorker,* October 23, 2023, https://www.newyorker.com.

22 Leta Hong Fincher analyses the campaign in *Leftover Women: The Resurgence of Gender Inequality in China* (London and New York: Zed Press, 2014).

been encouraging the establishment of more childcare centers, and, as an additional carrot, has raised its subsidies and maternity benefits for urban households contemplating a second or third child. Yet, as in other countries, beginning with Japan in the 1990s, where even stronger incentives have been offered, it has not made much of a difference. In the much-touted case of Hungary, more generous benefits did help to slightly boost fertility rates, but did not result in families having more than one or two children. No more persuasive is Xi Jinping's rhetoric about the need to uphold family values and restore women's role as essential caregivers.[23] One factor that might have helped is popular superstition. The Year of the Dragon is supposed to be a propitious one for births, and 2024's Dragon year did deliver a slight increase, though that may have had as much to do with the final relaxation of COVID restrictions.

When the birth rate drops well below replacement, it enters what demographers call a "low-fertility trap," from which there is little prospect of a rebound. The global fertility rate remains slightly above replacement level, but two-thirds of the world's population lives in countries with sub-replacement rates, and they are falling everywhere, even in predominantly Muslim and Catholic countries with a strong cultural devotion to large families. Anxieties about "underpopulation" take different regional forms. In wealthy European and North American societies, they are shaped by racial animus about whites becoming a minority as more and more immigrants fill the gaps in the labor force.[24] In China, these fears are primarily driven by

---

23 See Dong Yige, "Gender Awakening, Care Crisis, and Made-in-China Feminism," in *China from Below: Critical Analysis & Grassroots Activism*, ed. Ralf Ruckus et al. (Berlin: Gongchao, 2021), https://www.gongchao.org.

24 Apocalyptic right-wing fears about the "race suicide" of white people have a long history, dating from Madison Grant's *The Passing of the Great Race: Or, The Racial Basis of European History* (1916), Oswald Spengler's *The Decline of the West* (1918), and Lothrop Stoddard's *The Rising Tide of Color: The Threat Against White World-Supremacy* (1920), to Patrick Buchanan's *The Death of the West: How Dying Populations and Immigrant Invasions Imperil Our Country and Civilization* (New York: St. Martin's Griffin, 2001) and Renaud Camus' *You Will Not Replace Us!* (2018). See also Michael Teitelbaum and Jay Winter, *The Fear of Population Decline* (London: Academic Press, 1985); Ben Wattenberg, *Fewer: How the New Demography of Depopulation Will Shape Our Future* (London: Rowman and Littlefield, 2005).

calculations about economic growth, but they are also underpinned by complications arising from the CCP's longstanding interest in birth planning.

That interest had its origin in the CPC's general commitment to planning after 1949, and it culminated in China's monumental program of birth restriction, known as the one-child policy, which ran for thirty-three years, from 1979 to 2013, and was finally terminated in 2016.[25] This program was the last of the Mao era's great state-level directives. Though it was instituted three years after his death, Mao himself had moderated his views on the need for population control, which he hitherto considered to be a bourgeois restraint on the unfettered aspirations of the Chinese revolution. The reform era that followed was guided by the policy of uneven development (the coastal cities would get rich first), and it was accompanied by the use of family planning as a key component of social control. Select sectors of the economy were freed from government monopoly at the same time as the state established new and intrusive levels of authority over household life. These two components cannot be considered separately even though they were presented and rationalized as such.

In the mid-2000s, just as its prohibition on "excess" births was being questioned by demographers, the CPC launched another high-level initiative known as *shengtai wenming* [ecological civilization]. For the previous quarter century, developmentalism had been pursued at all costs, resulting in ruinous environmental consequences that threatened to compromise the path of further economic growth. As a result, the country was confronted with multiple ecological challenges: severe land degradation and erosion, desertification, water scarcity and flooding, chronic pollution, glacier retreat, and biodiversity loss as well as the onset of climate change impacts.[26] The new paradigm of shengtai wenming was officially introduced by Hu Jintao in 2007, and it became a cornerstone of *Xi Jinping Thought on Socialism with Chinese Characteristics for a New Era* after the new president came

---

25 Tyrene White, *China's Longest Campaign: Birth Planning in the People's Republic, 1949–2005* (Ithaca: Cornell University Press, 2006).

26 Jonathan Watts spotlights many of these problems in his engaging ecological travelogue, *When a Billion Chinese Jump: How China Will Save Mankind—Or Destroy It* (New York: Scribner, 2010).

to power in 2012. China's rulers wanted to signal a departure from the ecocide redolent of the old socialist bloc economies, including their own Maoist one. Henceforth, the country's fields, rivers, air, and forests would no longer be ravaged by the unbridled expansion of industry. Unlike the Western environmentalist movement, concern about despoliation in China long predated industrialization, and the new top-level dispensation was aimed at harmonizing with the country's rich cultural tradition of respect for, and aesthetic appreciation of, nature.

Despite this shift towards a greener society, the foot was still on the pedal. None of the material gains of the reform era were to be sacrificed; China was simply moving, with the air of inevitability fostered by traditional socialist doctrine, into a new mode of production. Beijing also saw an opportunity, in the international arena, to lead the way in adopting environmental sustainability as a formula for economic growth. Other countries had talked about it and had reverted. With a strong hand at the helm of its state, China would see it through and win accolades in doing so. In the 1990s, best-selling books like Lester Brown's *Who Will Feed China?* saw the Middle Kingdom as a source of global ecological collapse. Today's titles are more along the lines of *Will China Save the Planet?*[27]

Of course, there was also a pithy slogan, coined by Xi as early as 2005, to summarize the new direction: "Clean waters and green mountains are as valuable as gold and silver mountains." The "two mountains" theory [*liangshan lun*], as it was officially named, was an eco-developmentalist project that only a strong and confident socialist state could undertake, and it looked forward to future prosperity while gesturing toward the restorative harmony of the past. The mountain rhetoric resonated both with ancient cultural traditions and with the CPC's own formative Jinggangshan experience in the 1920s.

Xi first used the phrase on a visit to Yucun, a small village in Zhejiang province that had relied heavily on limestone mining. Badly polluted, the village and surrounding towns, including the port city of

---

27 Lester Brown, *Who Will Feed China? Wake-up Call for a Small Planet* (New York, W.W. Norton, 1995); Barbara Finamore, *Will China Save the Planet?* (London: Wiley, 2018).

Huzhou, were subsequently curated as a national model for ecological development, focused on bamboo cultivation, green tourism, and low-carbon conversion.[28] Huzhou was among the first cities in the country to introduce the Gross Ecosystem Product (GEP) accounting system, which incorporates sustainability into the growth performance targets used by local officials.[29]

Since "ecological civilization" is supposed to be an all-embracing philosophy of societal health, how is it impacted by the rapid decline in birth rates, sometimes phrased in alarmist terms as the latest "population bomb"? Anxiety about overpopulation was a central pillar of Western environmental thought and action from the 1960s through the 1980s, and it influenced the adoption of China's one-child policy. Where does the new "crisis" of underpopulation sit within the CPC's paradigm of green economic development?

First of all, it is not clear how, or even whether, the one-child policy contributed to the go-go decades of China's economic growth. Even without the prevented births, the country's "demographic dividend" of cheap labor, especially young female factory workers, was still abundantly available during the years when its coastal regions served as a low-wage production platform for the rest of the world. Now that we are in the era of the three-child policy, demography, once again, is playing an uncertain role in the emergent mode of green governance. Are the state's efforts to boost the diminishing birth rate at odds with its professed determination to limit ecological degradation? Is a population with more generational balance necessary for a sustainable society? How can the retro messaging about women's reproductive responsibilities be reconciled with any modern, socialist commitment to gender equity? My return visit to Shanghai was preoccupied with thinking through some of these questions.

---

28 For official state media accounts, see "Mountainous Village Revives Under Xi's Ecological Guidance," *Xinhua*, August 15, 2023, https://english.news.cn; and "Huzhou Recognized Nationally for Green Development," *Peoples' Daily*, August 19, 2022, http://regional.chinadaily.com.cn/huzhou/.
29 "National Assessment Report on Policy related to Natural Capital Accounting in China," Research Center for Eco-Environmental Sciences, Chinese Academy of Sciences (October 2019), https://seea.un.org.

ROBOTS WILL NOT LIE FLAT

In other countries where the population is rapidly aging, raising the statutory retirement age has been floated as an option to help balance the state's pension books. Not surprisingly, it is usually met with strong popular resistance. Until the policy was finally changed in September 2024, China had one of the world's lowest retirement ages—sixty for men, fifty-five for female office workers, and fifty for female blue-collar workers. The increase, to be phased in gradually, was to sixty-three, fifty-eight, and fifty-five years respectively. One of the reasons offered was that working for longer is a necessary response to the shrinking labor supply—the state has announced an estimated shortage of 30 million workers in the manufacturing sector alone. But according to Yuan Ren, a demographer who I interviewed at Fudan University's Institute for Population Studies, "the decrease in pension funds is really the main reason for raising the age." He dismisses the government's claim about a "labor shortage." It is "fake," he retorts, "unemployment is rising, we have 850 million people of working age, there is still a labor abundance, and there is no evidence to support a shortage at least until 2050." So what is the real reason for the anxiety about population loss? In addition to the concerns about the under-funded pension system, Ren attributes it to the "government's own interest in tax collection and GDP growth." For a party that wants to survive in the way its counterparts in the Soviet bloc did not, larger households delivering more revenue are quite simply seen as a form of insurance.

As a demographer—a profession at the heart of so much national turmoil since the 1960s—Ren is remarkably agnostic about the role of population in economic development. "People worry too much about population, in the past when it was too much, and now when it is shrinking." In his view, the state should focus more on providing services to support the choices of households and of women in particular.[30] "The government should support those who want to have

---

30 China spends less on services to its citizens, like healthcare and social security, than other countries with similar income levels. Only 6 percent of GDP is devoted to "individual consumption, a share far lower than the BRICS group of merging nations. Joe Leahy and Haohsiang Ko, "China Spending on Citizens Trails Peer Nations," *Financial Times*, February 22, 2025, https://www.ft.com.

more children," he said, "but it should also support those who choose not to." In the meantime, in recent months, a propaganda effort had been launched to present the graying of China as an economic opportunity rather than as an existential threat. An accountant in a Beijing ministry office estimated that the "silver economy" of goods and services for elderly care was worth $974 billion and was expected to more than quadruple by 2035.[31]

As in other countries facing a shrinking and graying population, an obvious remedy is to admit immigrants to fill the gaps in the labor force. But, like the hikes in retirement age, that policy has also run into resistance—mostly in the form of racist pushback in rich nations. In China, strong currents of ethnic chauvinism are also likely to limit a more open immigrant policy. In any event, the government is more focused on moving the economy up the skills ladder. Foreigners are more scarce in China than in other East Asian countries like Japan and South Korea, and even North Korea, and those who are issued work permits tend to be highly-skilled, and classed as "foreign experts," in keeping with the CPC directive of "encourage the top, control the middle, and limit the bottom."[32] More will surely be needed, since one of the causes of population shrinkage is a noticeable rise in the emigration of people with the resources to move, and the skills to make themselves employable elsewhere.[33]

Mechanization is an equally obvious solution. In the world's most robot-intensive economy, automated production already dominates large sectors of Chinese manufacturing, and the rapid deployment of AI is transforming them further. After a much-publicized spate of employee suicides in 2010, Foxconn, employer of 1.2 million Chinese workers, announced plans to steadily replace them with robots. By 2016, the company had replaced 60,000 members of its

---

31 "China to Bolster Silver Economy Amid Population Aging," *Xinhua*, March 11, 2024, http://english.scio.gov.cn.

32 "China Needs Foreign Workers. So Why Won't It Embrace Immigration?," *The Economist*, May 4, 2023, https://www.economist.com.

33 According to Xinmei Liu, "310,000 Chinese left the country in each of the past two years, a 62 percent increase from the earlier average of around 191,000 per year over the decade through 2019. "This Isn't the China I Remember," *The New York Times*, April 29, 2024, https://www.nytimes.com.

workforce in one factory alone.[34] Service robots are being introduced in the health and care sectors, where there is an especially acute shortage of workers for the rapidly aging population. Ke Ren, also a demographer at Fudan University, told me about her preferred solution for looking after seniors. "They cannot rely on their children or grandchildren anymore so they will just have to use robots as caregivers," she predicted. "Not comprehensive robots, who do everything, but ones that wash, clean, and cook." Chinese people, she assured me, "do not have a negative reaction to robots, and they are cheap and efficient sources of labor." Of course, she added, "they cannot currently provide the same emotional or psychological support," though AI innovators are hard at work on developing senior care robots that can offer authentic companionship as well as health management functions. There are already more than one hundred Chinese robotics startups specializing in smart technology for elderly care.[35]

Few would bet against the Chinese state's ability to fill any future labor shortages with machines, but there are no robotic solutions for climate change. AI propagandists tout the lower carbon footprint of their generative technologies but the computational power they depend on will require a staggering amount of electricity and water supply. If they are deployed to service profit-driven growth, these advanced technologies will only accelerate the climate emergency, and, if the most sustainable pathway forward is one of de-growth—which decouples economic expansion from carbon emissions—it is not one that is on the horizon for Beijing.[36] China may be adjusting to a lower annual GDP growth rate than the old 8 percent plus, but it is only entering the late infancy of its increasing economic reliance on mass consumerism. At the CPC's Twentieth National Congress in 2022, Xi Jinping laid out the plan to increase per capita GDP to the level of "medium-level developed countries" by 2035 and to make

---

34 Jane Wakefield, "Foxconn Replaces '60,000 Factory Workers with Robots,'" *BBC*, May 25, 2016, https://www.bbc.com.

35 Fan Yiying, "China's Solution for a Growing Senior Care Crisis: Millions of Robots," *Sixth Tone*, June 14, 2023, https://www.sixthtone.com.

36 Minqi Li, "Degrowing China—By Collapse, Redistribution, or Planning?" *Monthly Review* 75, no. 3 (July–August 2023), https://monthlyreview.org.

China a leading global power by 2050.[37] Achieving those goals with a rapidly aging society is already a dubious endeavor. To do so in the face of a multi-scalar ecological collapse brought on by climate change will be even more of a challenge.

For the humans who will still be in the workforce, the only way to reach those targets is to increase their labor productivity, and growth in that area is also declining.[38] China's annual working hours are already the longest in the world, exemplified by the punishing "996" schedule of tech industry employees—from nine in the morning to nine in the evening, six days a week. It's no wonder that disaffected youth have taken to "lying flat" [tang ping], the much-discussed trend of dropping out—by rejecting conventional expectations about buying an apartment and a car, entering into a marriage, raising a family, and consuming material goods. In that same spirit is their rejection of long working hours and the adoption of a low-energy lifestyle at odds with the hard sacrifice of the Chinese work ethic. Lying flat builds on two other widely observed memes: the consciously apathetic youth mentality of sang, and the anthropological concept of "involution," applied to a society where the grinding dominance of meaningless competition—over GPAs and GDP alike—has stalled any sense of optimism about the future.[39]

---

37 The declaration preceded Washington's announcement of strict export controls of semiconductor chips to China. This is not simply a trade war in the race for the next generation of technology; it is backed by an armada. Ever since Barack Obama's "pivot to Asia," the encirclement of China by US military bases has been relentless. But the expansion of sanctions and export controls under the Biden administration may only intensify Beijing's drive for self-reliance, especially in the development of advanced technology. Alex Palmer, "'An Act of War': Inside America's Silicon Blockade Against China," *The New York Times*, July 12, 2023, https://www.nytimes.com.

38 Loren Brandt et al., "China's Productivity Slowdown and Future Growth Potential," Policy Research Working Paper, 9298, Macroeconomics, Trade and Investment Global Practice, World Bank (June 2020), https://documents.worldbank.org/en/publication/documents-reports.

39 Zeng Yuli, "Turn Off, Drop Out: Why Young Chinese Are Abandoning Ambition," *Sixth Tone*, June 27, 2017, https://www.sixthtone.com; Wang Qianni and Ge Shifan, "How One Obscure Word Captures Urban China's Unhappiness," *Sixth Tone*, November 4, 2020, https://www.sixthtone.com.

Since its appearance in 2021, and viral uptake as a social media phenomenon, authorities have pushed back against "lying flat," and have banned promotional merchandise related to this passive-aggressive stance. But few could deny it is a coherent response to the "three mountains"—the steep costs of housing, education, and health—that young people have to scale in pursuit of middle-class well-being. The challenges are especially acute for a generation propelled into higher education in ever greater numbers only to find limited employment options on the other side, even for graduates of top-ranked universities.[40] Given the intense pressure of competing for success in the face of these odds, the additional expense of raising a child can seem beyond their reach.

Lying flat took hold during the COVID pandemic, and it was often compared to the so-called "Great Resignation" in the US, when sizable numbers of employees quit their jobs in response to wage stagnation and the poor quality of work life. The instability generated by the pandemic prompted a broad range of sentiment; from a rise in nihilism, amplified by the climate emergency, to the more liberating sense of deliverance from lifestyle norms and rules. But in China, an additional factor was the response to the state's own zero-COVID policies. The lockdowns, and the authoritarian measures that accompanied them, were especially traumatic for Shanghainese, who lived through arguably the most excessive forms of pandemic management in 2022. One of the lasting legacies has been an erosion of belief in the government's ability to deliver meaningful work and welfare.

One warm spring day, I went to meet Diwei Chen in a coworking space hosted inside Starbucks on the east side of People's Park. She had heard it was a place where "people go in order to pretend to work." A researcher of workers in the creative industries, she described the customers as laid-off tech employees who showed up "to stay in sync with the rhythms and hours of the jobs they used to do." Given the hectic pace of Shanghai's coffee revolution, Starbucks was no longer a cool place to be, she said, but this coworking space was an environment where users could be "surrounded by people

---

40 Diego Gullotta and Lili Lin, "Beyond 'Rising Tides' and 'Lying Flat': Emergent Cultural Practices Among Youth in Urban China," *Comparative Literature and Culture* 24, no. 1 (2022): 1–16.

being busy and would not feel left out." In her opinion, it was a "feel-good experience," especially for those in their mid-thirties, who were no longer employable in a tech sector struggling to survive during the economic downturn. "To avoid becoming invisible in the job market," they might soon be reduced, she predicted, to filling robot maintenance jobs, or laboring at data annotation to help train AI machine learning models.[41]

Whatever the customers were doing—and we did not investigate too much—they were not lying flat. Chen thought that, for younger people, lying flat was still a prevalent option, and an entirely "reasonable" response to a set of bad options. Those who were still employed had taken to "dressing ugly" in their workplaces out of disrespect for the poor quality of their jobs. Others, she said, were opting out of meaningless jobs, "temporarily if they had no economic cushion," and for good, if they could afford another kind of life. She herself had decided to leave the higher education sector, in part because of its stultifying hierarchy but also because of increasing state surveillance and censorship of academic thought and content. The day before, a fellow academic I met at Fudan University had told me that classrooms were now equipped with surveillance cameras that automatically transcribed lectures for party cadres to consult. Chen had been the first in her small Fujianese town to go to college, let alone earn a PhD, and so she admitted that there were quite a few people back home who would be "disappointed" by her choice. But she could see no alternative: "for those of us who were born in the '90s, the Chinese dream of middle-class stability has not only lost its charm, it's simply no longer attainable."

Chen asked me what it was like for the young tech employees I interviewed in the early 2000s for my book *Fast Boat to China*. I told her that they saw themselves as the first Chinese generation to be groomed for a middle-class life of homeownership and market consumerism, but they were also acutely aware of how the price of their labor defined their place in the global supply chain. They knew

41 This "digital underclass" is already being drawn from the ranks of vocational high schoolers at subminimum pay. Viola Zhou and Caiwei Chen, "China's AI Boom Depends on an Army of Exploited Student Interns," *Rest of World*, September 14, 2023, https://restofworld.org.

they were in a sweet spot, favored by foreign companies moving their operations to coastal China. Chen had come of age in a more precarious time, where the expectation of security and social mobility was much less widely shared. Homeownership "seemed out of reach," even with housing prices falling, and she herself "could not even think about having a child." Like all of the women I interviewed, she reported some pressure from her mother to start a family. "I'm thirty-one, after all, so my mother still thinks there's hope." But for her, the choice of having a child was a "decision about the future," and she was "only pessimistic about the future." So, too, there was a part of her that regarded reproducing "as a form of cooperation with the government and its goal of making us all productive."

No one else I interviewed was so forthright in their opposition as Chen, but all were dismissive in one way or another about the government pressure. "My friends and I often joke about having babies for the good of the country," said Changxin Li, a clerical employee who had one daughter but was hesitant about committing to another child. "The government wants us to create more GDP by being double-income earners, but won't give us the help we need to be parents." If "we're going to do it for China," she mused, "then China has to do much more for us." Jingbei Zhao, a high school teacher, told me that the party branch in her workplace had circulated a document with clear messaging: "If you are of the right age, then you should think seriously about having children." She rolled her eyes, and quipped, "The only way they might be successful is by heavily taxing single people." Zhao said she was unlikely to have children because "women like me just don't have to marry anymore to survive." Social media had persuaded her about the many downsides of maternity. "It keeps feeding me reasons for not doing it—and from what I see, it seems there are more unhappy families out there than happy ones." Besides, she added, there is no longer "any halo surrounding the idea of creating the next generation."

Qianlu Yang, an accountant for the branch of a Japanese firm, spoke more darkly about the "curse" that her generation had inherited: "The government made a terrible decision, and did not acknowledge it, so the curse was passed on to us." She herself had suffered as an only child, and "decided not to repeat that misery by having a child." But she described her outlook in more philosophical terms:

"If we cannot deal honestly with our past, we cannot dwell on the future—the curse has left us with too much depression and anxiety." Like most of the women I interviewed, she subscribed to the general feeling that, even though there were less Chinese in the general population every year, there were still too many people—*ren tai duo*—competing for too few resources or opportunities. However widely shared, this belief is largely the legacy of the last few decades of intense attention to population control.

REN TAI DUO

By selecting *Adam Smith in Beijing* as the title of his well-regarded 2007 book, Giovanni Arrighi neatly captured readers' interest in the career of China's modern experiment in market economics. Yet it was another eighteenth-century British economist, Thomas Malthus, who arguably exerted more influence on China, and for much longer. His strenuous pronouncements about the relationship between population and food supply, and the nature of checks upon population growth in particular, attracted opprobrium and praise alike, feeding into twentieth-century state policy that took turns at being pro-natalist and anti-natalist.

Like Smith before him, Malthus had a good deal to say about China in his 1798 *Essay on the Principle of Population*, singling out, as an object of analysis, the seeming contradiction between the country's great wealth and high soil fertility on the one hand, and its "stationary" social condition on the other. As far as empirical knowledge about China went, he did not have much to go on: some information from Jesuit missionaries, travelers' accounts, and observations from members of the ill-fated Macartney Mission to the Qianlong emperor in 1793. Malthus was not alone in his professed amazement at the population estimate of 333 million brought back by George Staunton, a Chinese-speaking member of that mission.

Malthus conceded that "a minute history of the customs of the lower Chinese would be of the greatest use in ascertaining in what manner the checks to a further population operate; what are the vices, and what are the distresses that prevent an increase of numbers beyond the ability of the country to support." But the absence of this detailed information did not prevent him from speculating about these checks, vices, and distresses. Indeed, China proved to be an important foil for Malthus, allowing him to present a rigid comparison between East and West. In his view, curbs on excessive population growth could take two forms: a preventive or a positive check. The former he considered to be the superior European mode, and it was characterized by the "moral" restraint exercised by virtuous Christian individuals in choosing to delay marriage until they could support a family. Rationalized family planning along these lines was presumed to be non-existent in non-Western societies like China, where too many people married too young, resulting in unrestrained reproductive activity, and where the population checks therefore had to come in the form of war, famine, or pestilence. For Malthus, the ability to assess the costs and benefits of starting a family accounted for the general prosperity of middle-class Western households, whereas the inability to delay or abstain from having children accounted for the misery of the bulk of the Chinese people. As a result, their poverty had everything to do with the prospect of having 333 million mouths to feed and also with the downward wage pressure mandated by such a vast labor supply.

These blunt generalizations, feeding into an equally binary view of East and West were not untypical of Orientalist thought, past and present. Indeed, by the time he took to revising his essay (the second version was not completed until 1826 and contains a whole chapter on China), Malthus had taken a teaching post as professor of history and political economy at the East India Company's college in Hertfordshire. His students, who were Orientalists-in-training for managerial positions in the Far East, referred to him as "Population" Malthus. The records that survive of his teaching suggest that Smith's *The Wealth of Nations* was his core curricular text, and that a detailed knowledge of Indian political economy was not a requirement, but it is safe to assume that he nonetheless imparted lessons from his own *Essay* about the "positive checks" of famine and war on population.[42] One can also presume that the company's own role in creating these conditions of famine and war went unacknowledged. Nor is it irrel-

---

42 Keith Tribe, "Professors Malthus and Jones: Political Economy at the East India College 1806–1858," *The European Journal of the History of Economic Thought* 2, no. 2 (1995): 327–54.

evant that the firm had been selling opium to Chinese merchants for several decades as a way to balance its lucrative tea trade.[43] Malthus' employer had a vested interest in the "opening up" of China to Western trade. After the cruel Opium Wars, the subsequent career of that trade was aggressively administered by people who upheld an Orientalist mentality in all of their dealings with the Chinese.

Malthus' conclusions about the undesirable Chinese combination of high fertility and high mortality would have been integral to that mentality, because they offered a long-abiding stock explanation for poverty and misery in the Middle Kingdom. These assumptions finally fell apart when demographers were able to access better data. Demographic historians James Lee and Wang Feng delivered the most stinging rebuke in 2001 when they offered an alternative view of population restraints in Imperial China. They found that female infanticide, in particular, was widely practiced as a way to control family growth, reflecting a longstanding prejudice against daughters. This form of discrimination, in turn, affected the marriage market. Girls in the numerical minority married early, but the lack of brides consigned large numbers of males to bachelorhood and celibacy. More striking, Lee and Feng found much lower levels of fertility than Malthus had assumed, almost as low as the European equivalents. Last but not least were the high rates of adoption, or fictive kinship, cited by Lee and Feng—"at least one of every 100 Chinese children in the past." The strong sense of "entitlement to children" meant that nuptiality was not the only determinant of family formation, and that widespread adoption of both sons and daughters provided lots of flexibility.[44]

In place of Malthus' picture of the Chinese incapacity for individual restraint, Lee and Feng saw the hand of collective regulation in maintaining the balance of population and resources. It was not "vice and misery" or environmental stress that determined the occurrence of births, but rational and often complex decision-making at

---

43 See Alison Bashford and Joyce Chaplin, *The New Worlds of Thomas Robert Malthus: Re-Reading the Principle of Population* (Princeton: Princeton University Press, 2016).

44 James Lee and Wang Feng, *One Quarter of Humanity: Malthusian Mythology and Chinese Realities, 1700–2000* (Cambridge, MA: Harvard University Press, 2001).

the family and community level. There was a distinction between East and West, they concluded, but it was between choices made out of individual self-interest and those based on calculations about the communal good. Over time, and especially in the course of the late twentieth century, these collective judgements would be transferred from the extended family to the state.

Lee and Feng's revisions came at the end of a century when the influence of Malthus on Chinese population policy had swung like a pendulum. The republican revolutionary Sun Yat-sen, for example, was sharply critical of the neo-Malthusian group of Social Darwinists (Herbert Spencer, William Graham Sumner, Francis Galton) for attributing poverty to overpopulation. Instead, he blamed the colonial stranglehold over China's economy. His nationalist successors, like Chiang Kai-shek, adopted a stronger Confucian ideology, and, under the influence of their American allies, embraced some of the eugenicist thinking of the second wave of neo-Malthusians (Fairfield Osborn, Lothrop Stoddard, Madison Grant) in the 1920s and 1930s. When the CPC swept to power in 1949, Malthus was reestablished as an archenemy of the people. This had been Marx's view after all, for whom poverty was a result of the unequal distribution of wealth, not overpopulation. Just as the aim of the first edition of Malthus' *Essay* had been to argue *against* "the future improvement of society," (skewering Rousseauists such as William Godwin and the Marquis de Condorcet) the chief thrust of the second edition was to condemn England's Poor Laws. In the view of its author, any effort to provide relief to working people was a waste of time and resources since it would simply result in even more mouths to feed. This conclusion was presented as a natural law, according to Malthus' bogus proposition that population multiplies geometrically and food arithmetically.

By contrast, the "improvement of society" espoused by Mao was to be conducted through "Promethean" means. Now that the Chinese people had "stood up," there were no limits, natural or otherwise, to their potential to develop. Population growth, according to Mao's credo of "more people, more power," was an unconditional good. Even so, officials were stunned when the first census taken of the Chinese people in 1953 showed about 583 million residing in the mainland alone. With mortality in sharp decline as a result of social gains and absence of wars, that number would almost double by the end of the century. The first serious challenge to Mao's pro-natalism came from the economist Ma Yinchu, who presented his New Population Theory at the First National People's Congress in 1957. Arguing that rapid population growth would hinder China's development, he boldly advocated for birth controls. Ma Yinchu was disgraced for his views (and was not rehabilitated until 1979), but Mao nonetheless saw an opportunity to extend the wisdom of state planning into the realm of population. By the early 1970s, food supply had tightened, and so the rhetoric of birth planning was introduced through slogans like "later, longer, fewer" [wan-xi-shao] or "one child is not too few, two is just enough, and three is too many." These "soft" policies appeared to result in notable drops in birth rates.

By then, the Western development establishment was in the vice-grip of a third Neo-Malthusian wave (William Vogt, Paul and Anne Ehrlich, Lester Brown) that regarded population overshoot as the root cause of impending mass starvation, if not outright ecological collapse. Excessive breeding on the part of poor, Brown people in the Third World was to blame, and had to be curtailed.[45] The gospel of overpopulationism, heavily promoted by the Rockefeller, and later, the Ford Foundation, culminated in the massive forced sterilization program launched by India's government in the mid-1970s, along with the adoption of strict measures in many of the other twenty or more countries that took the offer of Western aid in return for commitments to cut their birth rates.[46]

One of the political drivers of this reactionary wave was the belief, embraced within the Rockefeller Foundation, that overpop-

45 Matthew Connelly places these events in a broad historical perspective in *Fatal Misconception: The Struggle to Control World Population* (Cambridge, MA: Harvard University Press, 2008). Also see Betsy Hartmann's classic feminist survey, *Reproductive Rights and Wrongs* (New York: Harper & Row, 1987). For a slightly updated version of overpopulationism, see Lester R. Brown, Gary Gardner, and Brian Halweil, *Beyond Malthus: The Nineteen Dimensions of the Population Challenge* (New York: Routledge, 1999).

46 See Mahmood Mamdani's influential critique of the population control crusade in *The Myth of Population Control: Family, Caste and Class in an Indian Village* (New York: Monthly Review Press, 1972).

ulation was a chief cause of the Chinese revolution, and that birth control programs could help stave off the advance of communism elsewhere.[47] CPC leaders were rightfully wary of the counterrevolutionary underpinnings of these campaigns, but also mistrustful of advice from their own social scientists. Instead, they looked to the imprimatur of hard science to incorporate a stricter version of population control into their policymaking. Bizarrely, they leaned on the cybernetics studies of a missile engineer, Song Jian, to reach their decisions about what would become known as the one-child policy. CPC leaders gave fuller credence to Song's mathematical estimates and his approach to population growth as a purely quantitative or biological problem.[48] Swayed by the modeling deployed in the Club of Rome's seminal 1972 *Limits to Growth* report on the impending ecological crisis, they also took seriously Song's conclusion that the optimal population for which China had a carrying capacity should be around 650–700 million people; these numbers had already been surpassed by almost a third. After internal communications in 1979, an Open Letter encouraging one-child families was issued by the National People's Congress in September 1980, inaugurating one of modernity's most comprehensive experiments in social engineering. Soon, birth-planning slogans were everywhere—on walls, media communiques, and in character posters. The Department of Family Planning, upgraded to a top-level agency with as many as half a million local branches, began to mushroom in size (an estimated twenty million were members at its peak) as the campaign to reduce childbearing took on the character of a mass mobilization.

Although birth rates had been declining sharply since the mid-1960s (and even earlier in Shanghai, where contraception was more widely available in the 1950s), policymakers reached a new consensus that population growth was a primary cause of poverty and therefore an obstacle in the path of China's goal of raising per cap-

---

47 John Bellamy Foster, "Malthus' Essay on Population at Age 200: A Marxian View," *Monthly Review* 50, no. 7 (December 1998), https://monthlyreview.org.
48 Susan Greenhalgh offers a comprehensive analysis of why Song's empirical computations were preferred over the social scientists in the Peoples' University's population studies group in *Just One Child: Science and Policy in Deng's China* (Berkeley: University of California Press, 2008).

ita living standards to match those of developed countries. There was no evidence to support this neo-Malthusian proposition that smaller families would necessarily lead to wealth, but it became a cornerstone of Deng Xiaoping's reformist policies. As a reward for minimizing births, the government promised free access to birth control supplies and abortions, paid vacations, health care subsidies, salary supplements, guaranteed retirement income, and privileged housing opportunities. By the early 1980s, the planning authorities began to introduce penalties, including mandatory abortion, IUD insertion, sterilization, demotion and other financial disincentives. Officials kept close tabs on women's fertile periods, and pregnancies had to be authorized. The consequences of these draconian policies were far-reaching. Infanticide returned, abandonment of babies, especially females, proliferated, and overly harsh punishments for "illegal seconds" exposed the cruelty of minor bureaucrats. The policing of women's bodies and minds, especially in rural areas, was a grisly intervention into the most intimate sanctum of China's family-based cultures.

In Western countries, the excessive zeal with which local Chinese officials implemented the policy was seized upon as confirmation of the inhumanity of communist state power. Yet for all of the tragic stories that circulated, the evidence of ethnographers like Susan Greenhalgh was that many Chinese women saw the advantages of the policy, not because it benefited the national good, but because it served their own purposes.

> I met many women, urban and rural, who were grateful for the birth limits. I will never forget the smile on the face of the Shaanxi villager who in 1986 sought me out to tell me proudly that she was the very first to be sterilized, and so able to avoid more unwanted pregnancies. Or the delight of the professional women in Beijing who in 1999 praised the policy for enabling them to dodge their mother-in-laws' demands for more children and thus have the freedom to develop their careers.[49]

---

49 Susan Greenhalgh, "Same Old Coercion Story," in "What a Picture of China's One-Child Policy Leaves Out," *China File*, February 6, 2020, https://www.chinafile.com.

Even so, Greenhalgh concedes that the policy "profoundly remade China and its people," shaping their capacity to compete in modern labor and consumer markets.[50] It also offered the CPC a new kind of reach over its population; through what Michel Foucault defined as *biopolitical* power over the production and cultivation of life itself.[51]

In any event, the one-child policy was not evenly implemented. After a good deal of pushback following the traumatic program of sterilizations in 1983, rural households were generally allowed to follow a de facto two-child policy, while many families had unauthorized children or simply paid the excess-child fines (lucrative for local officials) for each out-of-plan baby.[52] By some estimates, only a third of the population—ethnically Han, urban, and employed by the state— were restricted to one child. By the 1990s, and after a torrent of international condemnation, the most coercive measures were relaxed.

In East Asian countries without population controls, fertility levels declined at the same rate as in China, suggesting that economic development led to smaller families, and not the opposite, as the CPC had assumed. Yet, notwithstanding the probability of mass underreported births—the so-called "missing girls"—there is every reason to believe that the policy did hasten the decline in Chinese births. The official line that it prevented 400 million births has been disputed however, with some demographers suggesting only half of that figure. More serious is the charge that officials failed to anticipate the consequences of *population shrinkage*, and that, despite advice from demographers and other social scientists, they pulled the controls back too slowly for the plummeting numbers to recover. Given how widely government statistics were distrusted, especially the data produced by the family planning commission to promote the success of its policy, it is quite possible that cadres themselves did not believe

---

50 Susan Greenhalgh, "Why Does the End of the One-Child Policy Matter?" in *The China Questions: Critical Insights into a Rising Power*, ed. Jennifer Rudolph and Michael Szonyi (Cambridge, MA: Harvard University Press, 2018), 184.

51 Susan Greenhalgh and Edwin A. Winckler, *Governing China's Population: From Leninist to Neoliberal Biopolitics* (Stanford: Stanford University Press, 2005).

52 When provinces and cities were asked to record their receipts of fines in 2013, well after the heyday of these excess-child penalties, the two-thirds who did so declared revenue amounting to $3 billion. Beardson, *Ageing Giant*, 13.

the birth rates were falling so fast. It was not until 2010 that the policy was significantly weakened. By then, free contraception and abortion facilities were widely available, and women had increasing access to higher education. As numerous studies have shown, easy access to birth control and female education are the two factors considered most critical, worldwide, to a decline in birth rates.

Even so, the newly available choices still produced skewed gender ratios. China led the world in gender disparity in the 2000s, with some estimates of 121 boys to 100 girls (the "natural" ratio is 105 to 100). Access to portable ultrasound machines (GE's best-selling import) enabled sex-selection, and even after they were banned in 2002, blood test results were still available through the cross-border trade with labs in Hong Kong and Macau. Orientalists have long condemned female infanticide as evidence of a "barbaric" culture, and yet, as Mara Hvistendahl points out, the abortion of female fetuses after prenatal screenings is often seen, hypocritically, as the mark of an "advanced" society.[53] Notably, recorded gender imbalances were greatest in Asian countries that saw the largest drops in births.[54] As single-child families have become a normative, optional choice over the last decade, the gender ratios have improved, but studies show that the skewing is greater for second and third births—a trend that does not augur well for societies like China, that are now leaning on women to procreate more.

The demographic quagmire that emerged from the one-child policy was not only that of a shrinking population, but one in which the "bare branches" were unable to compete for less numerous, available female mates and were consigned to social disrepute as a result. I found the desperation on display during a visit to Shanghai's "matchmaking market" in People's Park. Once a week, several hundred parents and grandparents lay out written descriptions of their offspring in hopes of attracting interest. The profiles of the daughters and

---

53 Mara Hvistendahl, *Unnatural Selection: Choosing Boys Over Girls, and the Consequences of a World Full of Men* (New York: PublicAffairs, 2012), 65.

54 Hvistendahl, *Unnatural Selection*, 43. Hvistendahl cites evidence that similar sex ratios could be found among Asian Americans, of Korean, Indian, and Chinese descent. Over time, Asian Americans have continued to show the lowest total fertility rate among all population groups in the US.

granddaughters dwell on their physique and aesthetic appeal, while leaving out their educational achievements, presumably because any mention of college degrees will put off potential male suitors who are fearful of falling below expectations.

One of my interviewees, Shufu Zhang, described the market as a "throwback" to an earlier era, but also as a reflection of the state's anxiety about social reproduction. The subsidies in healthcare, income, and housing that had once been used to reward parents of single children had proven less effective when they were offered to parents willing to have up to three children, nor were they as widely available. Recently married, Zhang was witnessing firsthand how "women like me are extra sensitive to any talk from the government about reproduction." "It's one thing to talk about 'encouraging' us to have children, because they need more workers," she said, "but I wonder what difference it would make if they actually provided real support like much longer parental leave and early childcare and other economic incentives?"

For her, the experience of being an only child was a positive one. "We got a lot of loving attention," she admitted, "and since everyone else was single, we never saw that as abnormal—our cousins were basically our siblings." Out of necessity, they "made kin" and built an extended, or alternative kind of family, as Donna Haraway and others have described it.[55] I spoke with interviewees who described the unacceptable cost of being an only child—overbearing parents who closely monitored their food intake, drove them too hard to perform academically, and restricted their movements outside the home to ensure their safety at all times. As a result, food neuroses were not uncommon, and at least two of my interviewees partly attributed their reluctance to marry to the parenting regimen they had endured. Huiying Wang, a manager in an insurance company, confided that "my father has left such a bad mark on my mind that I don't think I can trust a man to share my life with."

---

55 Donna Haraway and Adele Clarke (eds.), *Making Kin Not Population* (Chicago: Prickly Paradigm Press, 2018).

Other than the mixed blessing of their exposure to "over-caring," what they also had in common were parents and grandparents whose education was truncated by the Cultural Revolution, and who had to settle for less in their lives. All the more reason why the elders looked to their offspring for forms of support. Wang was already feeling the pressure of looking out for her grandparents, who live a hundred miles inland. "It won't be long before my parents retire, and I will be in the 1-2-4 hole—just one child to look after two parents and four grandparents. They don't want to be a burden on us, but I know it will not be easy for me."

Baohua Chen had just returned from a decade working in Beijing in order to be close to her recently retired parents. "I feel that I owe them," she said, "they gave so much to me." Besides, she observed, we are "in a very dangerous time; the economy is bad, the regime is more repressive, and I fear that military conflict is coming." Like others I interviewed, she thought that people's views about the government had radically changed as a result of the zero-COVID lockdown. "No one believed they could treat us this way, especially in Shanghai," she explained, "and even I, who have low expectations of officials, was shocked." Chinese people, she said, are very practical—"they can adapt to changing political conditions"—and the only real constant is "family values—the family is the most important thing." That said, she had no interest in starting one of her own. "I have never really liked children," she admitted, "and so I tend to avoid social gatherings with friends who do have families."

Xiaoyan Li, one of Chen's recently married college friends, said that she intends to have children soon, and more than one. But she was dismissive of the view that having children these days is a mark of status. As we spoke, she rolled her eyes at the mention of upper middle-class friends who are already parents and who worry about the costs of a second child. Their "concerns are real," she conceded, but she finds it odd that they "often talk about children as if they are expensive cars, as something you can buy and keep in good condition but only if you have enough money." For her, the "lesson of modern economic life in China" is that people "just need to live within their means." And, she added, "so does this country."

QUALITY UPGRADE

In 1990, environmental scientists Qu Geping and Li Jinchang published *Population and the Environment in China*, a seminal book that summarized the ecological toll of China's burgeoning population on soil quality, arable productivity, deforestation, water shortages, pollution, energy supply, and general living standards. Qu and Li concluded that the causal relationship between overpopulation and environmental degradation was a grave threat to the country's development. In fact, "population control and environmental protection," they argued, "are two sides of the same coin."[56]

Hitching environmental protection to population control at the height of the one-child policy was a shrewd move to promote the cause. But Qu and Li's book also reinforced the idea that overpopulation was more responsible for the ruinous environmental damage than state policies of unchecked industrialization. After all, how many forests were denuded to fuel the backyard steel furnaces decreed during the Great Leap Forward? How many environmental regulations were waived to attract foreign investors to the special economic zones during the first wave of offshore outsourcing in the 1980s? These were delicate matters to broach for well-placed commentators, hoping to influence state policy on the environment without being directly critical of CPC decision-makers. In any event, Qu walked the line successfully and went on to become the first, and long serving, director of China's Environmental Protection Agency, shaping the policy that would result in the adoption of "ecological civilization" as a new framework for sustainable development.

Despite the ending of China's one-child policy, the legacy of population determinism lives on in the popular mind—fewer people must surely mean less environmental stress and damage. At the same time, the government is promoting its de facto three-child policy in a bid to offset the graying of China. This contradiction lies at the heart of shengtai wenming and is more or less unresolvable. When I met him in Shanghai, Yifei Li, the coauthor of a persuasive book

---

56 Qu Geping and Li Jinchang, *Population and the Environment in China*, trans. Baozhong Jiang and Ran Gu (Boulder: Lynne Rienner, 1994), 7.

about China's heavy-handed application of green governance (which he terms "environmental authoritarianism"), confessed that he had neglected to see the connection between population and environmental policy.[57] "I was surprised when I got your email request to meet," he told me. After all, these fields tend to be separated in the academy, he said, and in the state bureaucracy alike. For Li, the oversight may have been due to his age as a younger scholar, and for the bureaucracy, a reflection of the need to forget a former policy widely seen as an example of state overreach.

But perhaps there is an opportunity to be seized in this separation of spheres. After more than a half century in the driver's seat, it may be time to minimize the population factor in debates about sustainability. Anti-Malthusian environmentalists have long argued that the quantity of people is much less important than how they consume goods and energy. Beginning with their intervention at the 1994 International Conference on Population and Development (also known as the Cairo Conference), feminists have pushed back against the coercive calculus of fertility control and established that women's environmental well-being is best served by increasing access to sexual and reproductive healthcare, income, gender equality, and employment.

The urgency of confronting the climate crisis has brought back the populationist habit of fearfully eying the environmental impact of human numbers. Birth rates may be declining but the overall numbers are still growing. Surely, we are told, population pressure on the Earth and atmospheric carbon cannot be ignored; it is time to break the taboo on population engineering and reconsider these impacts in the more pressing light of the climate emergency. Once again, we see the temptation to pivot back toward the womb as the focus of attention and sidestep the reality that fossil fuel extraction and high-carbon overconsumption are root causes of CO_2 emissions. Populationists who have absorbed the feminist lessons are reissuing their siren song. This time, their pathway for solutions to climate change

---

57 Yifei Li and Judith Shapiro, *China Goes Green: Coercive Environmentalism for a Troubled Planet* (Cambridge: Polity Press, 2020).

lies through respect for reproductive justice, women's empowerment, and voluntary choice but it is still focused on reducing childbearing.[58]

Is this newly packaged approach to fertility reflected anywhere in China's policy on climate change? Li, who is a keen observer of the workings of bureaucracy, was skeptical: "The government is even reluctant to attribute weird weather events—of which there are increasingly many—to the impact of climate change." He was not even sure there is anything like a climate policy plan. According to Li, this reluctance is not due to "climate denial," but is a result of "conservatism" in the face of a problem that may be beyond the ability of a centrally planned state to fix.[59]

That has not stopped officials from making strong interventions at international meetings of climate policymakers. At the 2007 UNFCCC conference in Vienna, China's senior climate negotiator, Su Wei, made the audacious claim that his country's population controls had prevented 300 million births and it had therefore saved 1.3 billion tonnes of carbon dioxide emissions a year. The implication, which was ridiculed by representatives from other countries, was that China should receive annual credit in its emissions budget for these unborn people, as well as retrospective credit for more than two decades of restricted population growth. It was a strategic, if misguided, move in the game of carbon calculus that climate negotiators like Su Wei would learn to play in the years to follow as they aimed to avoid restrictions on their country's emissions.

With more success, China has long maintained that it should be designated as a developing country. In accordance with the Kyoto doctrine of "common but differentiated responsibilities," this status makes it less subject to the stringent emissions reductions laid out

58 Jade Sasser analyzes this tendency in *On Infertile Ground: Population Control and Women's Rights in the Era of Climate Change* (New York: New York University Press, 2018).

59 Despite the CPC's long record of calling on individuals to sacrifice for the national good, a recently introduced personal carbon accounting scheme—whereby consumers can earn favorable financial products and services in return for their low-carbon behavior—is an incentivized individualistic program. Jiang Yifan, "Shanghai to Trial Personal Carbon Accounts," *Dialogue Earth*, July 7, 2022, https://dialogue.earth/en/.

in the UNFCCC agreements. For that reason, it aligns itself with the Global South nations who have contributed much less to climate change over the last few centuries than the countries who were beneficiaries of high-carbon industrialization. In common with them, Beijing argues that it is up to the Northern nations to make the sacrifices first, while those who are less responsible for climate change should be given time to develop their economies before they can assume the burden of cutting emissions. According to this view, China is among those who are owed a climate debt by the rich nations for as much as half a millennium of colonial plunder (albeit only a century, in China's case). By that same token, Beijing should not be on the hook to make climate finance payments into the UNFCCC funds earmarked for low and middle-income countries to help them reduce emissions and adapt to climate impacts. Given the rapid rate of China's development, this position has become more and more untenable, but it still provides a rationale for sidestepping the most binding kinds of burden-sharing pledges.

After facing heavy criticism for its part in undermining a strong action plan at Copenhagen, Beijing made amends by playing a costarring role in sealing the landmark Paris agreement in 2015. Yet it has struck a cautious tone ever since, especially when it comes to protecting its reliance on coal, the dirtiest of all fossil fuels, and currently accounting for up to 60 percent of its domestic energy consumption.[60] At Glasgow's COP26, the Chinese delegation insisted that the final text called for a "phase down" instead of "phase out" of unabated coal power. And at COP28, in Dubai, it lobbied against phase out, and in favor of the final text that called for the much softer "transition away from fossil fuel."

In the run-up to COP28, China confirmed its pledge to reduce emissions from peak carbon by 2030 and to achieve carbon neutrality by 2060. Initial estimates from NGO trackers, like the Centre for Research on Energy and Clean Air, Global Energy Monitor, and Carbon Brief, showed that the country was off track in its effort to meet those commitments. But the domestic increase in hydro and

---

60 Greenpeace, "China Approves Coal Surge Despite Emissions Pledge," *Al Jazeera*, April 24, 2023, https://www.aljazeera.com.

renewable energy installations has been so rapid that their forecasts had to be revised; some suggested that emissions would now peak in 2025, with a decline thereafter.[61] China's output and installation of clean technology towers over all other national or regional efforts; it is building two-thirds of all the global solar and wind plants, it already generates 80 percent of the world's solar capacity, to be transmitted to all corners of the country by 2030 through an ultra-modernized grid, and it also supplies more than half of the Lithium-ion batteries, and a huge volume of rare earth elements to other countries.[62] But it has also become the world's largest carbon emitter, accounting for more than a quarter of global emissions, and it is the world's largest importer of gas and oil. As China emerged from the lockdown, it issued permits for six times the number of coal-driven power stations as the rest of the world combined; the equivalent of two new plants per week.[63] Permits increased fourfold from 2022 to 2023.[64] Paradoxically, the surge in permits was prompted by climate change impacts; a decrease in available hydropower from drought-beset rivers, and soaring energy demand to combat the heatwaves of 2022.

Elsewhere in the world, Global Coal Plant Tracker showed record *low* levels of new plant construction while China accounted for 95 percent of the new construction.[65] The country's continued reliance on coal power as an energy staple, accounting for 70 percent of national emissions, appears to undercut any professed goals of

---

61 Lauri Myllyvirta, "China's Emissions Set to Fall in 2024 After Record Growth in Clean Energy," *Carbon Brief*, November 13, 2023, https://www.carbonbrief.org.

62 Benjamin Schuetze, "The Uneven Politics of Decarbonization in the Middle East and North Africa," *Middle East Report* 311 (Summer 2024), https://merip.org.

63 Julia Simon, "China is Building Six Times More New Coal Plants Than Other Countries," *NPR*, March 2, 2023, https://www.npr.org.

64 Flora Champenois, Lauri Myllyvirta, Qi Qin, and Xing Zhang, "China Off Track on All Key Climate Commitments as Coal Power Approvals Continue," *Global Energy Monitor* (February 2024), https://globalenergymonitor.org.

65 According to the Global Energy Monitor's Global Coal Plant Tracker, https://globalenergymonitor.org/projects/global-coal-plant-tracker/. However, in the first half of 2024, the number of permits approved fell considerably. Fiona Harvey and Amy Hawkins, "China's Coal-Fired Power Boom May be Ending Amid Slowdown in Permits," *The Guardian*, August 22, 2024, https://www.theguardian.com.

carbon reductions.[66] India's fast growth is also coal-dependent, but it still lags far behind China, which consumes and produces more than half of the world's coal, while leading the world as the largest importer, accounting for up to one-third of the global coal trade.[67] It is also the world's largest exporter of metallurgical coke (used in the production of steel), and was the largest public financier of overseas coal plants, in countries as diverse as Vietnam, Pakistan, South Africa and Bosnia, until it pulled out of the business in 2021. These statistics are staggering, and they put a serious dent in the impetus of shengtai wenming. Notably, central policymakers who live in the key cities are less impacted personally by the pollution generated by the plants. Beijing shuttered its last coal plant in 2017, amid a rash of closures of dirty industry in and around major cities and has reduced the number of its notorious smog days as a result. Shanghai's air is much cleaner than when I lived there in 2002, but many other provincial cities still fall well below ambient air quality standards.

In the US, race is a decisive factor in the uneven distribution of airborne pollution. One recent study showed that 78 percent of African Americans lived within thirty miles of a coal-fired power plant. This is the kind of statistic that jump-started the environmental justice movement in the late 1980s. By contrast, the spatial distribution of China's plants and dirty industry is not race-based, but it does show clear disparities between eastern regions and the western ones that bear a greater environmental burden, and even more so between cities and rural areas. More environmental benefits accrue to the higher-income populations in urban centers. To secure the blue skies of key cities, many workers employed in the polluting factories in surrounding localities lost their jobs and have not benefited from

---

66 Coal-fired plants are a major source of emissions. Richard Smith points out, "most of the rest come from what are called the 'hard-to-abate' industries like steel, aluminum, cement, aviation, shipping, chemicals, plastics, textiles, and electronics. They cannot be significantly abated either in the United States or in China with current or foreseeable technologies." "China's Engine of Environmental Collapse," in *China from Below*, 45, https://www.gongchao.org/en/china-from-below/.

67 "Coal 2023: Analysis and Forecast to 2026," International Energy Agency (2023), https://www.iea.org/reports/coal-2023/executive-summary/.

the "just transition" to new ones that is a fundamental principle of climate justice.[68]

Nationally, the vast majority of China's absolute poor population are located in ecologically fragile areas—some estimates suggest about 200 million people living in ecological poverty, susceptible to increasingly severe climate impact extremes. But relatively affluent coastal regions are hardly exempt. They are also subject to floods from torrential rains and typhoons, life-threatening heatwaves, and evaporation of surface water supplies. Glacier retreat and desertification affect some regions more than others, but there are fewer and fewer locations that are not ecologically vulnerable. Which population groups are being protected in these places? How a locality builds adaptation measures usually offers the most telling evidence of whether principles of climate justice are being applied.

On my return visit to Shanghai, I lived near NYU Shanghai's campus in Qiantan, a district bordering the eastern bank of the Huangpu River that has been redeveloped with upper middle-class residents in mind. Conceived as "the New Bund," with a world financial center, upscale high-rise apartment complexes, and ultra-luxury shopping malls, it is the kind of massive urban investment that might have made sense in the more upbeat pre-COVID economy. Qiantan's consumer showcase is the landmark retail complex of Taikoo Li, a wellness-themed "experiential mall" whose top-brand stores are cocooned in a 120,000-square-meter environment blessed with sensitive light and vegetation features, and an AI-digitalized running track called "Sky Loop." The shop assistants surely have one of the loneliest jobs in the city, since customers are few and far between.

The district's showpiece is a well-curated chunk of the regenerated Huangpu waterfront that has been tastefully landscaped with leafy parks, broad promenades, hiking trails, biking paths, and cultural amenities that include high-end arts centers. This waterfront corridor was planned and constructed in less than two years, in accord with the city's *Three-Year Plan for Public Space Construction on Both Banks of the Huangpu River (2015–2017)* and as part of the munic-

---

68 Yaqiu Wang, "Beijing's Green Fist," *Human Rights Watch*, March 29, 2022, https://www.hrw.org.

ipal government's effort to make Shanghai "an excellent global city." The shoreline connection project involved the demolition of multiple industrial facilities and residential neighborhoods. Completed in 2017, the riverside greenways between the Nanpu Bridge and Lupu Bridge are now a vast public amenity, conceived as an environmental and recreational benefit for the people of Shanghai.[69] Yet, as Yifei Li and Xiaohua Zhong discovered, consultation about what the "people" might actually want from the process of "returning the river to the people" was minimal. A survey was distributed by the state-owned development company, but it transpired that the city had already approved the plan before the results were evaluated. As one official put it, "listening to the people is more of a formality. Objectively speaking, public opinions cannot play a decisive role in the plan."[70] Nor was there much choice for the thousands of residents who were displaced by the development. Their homes and neighborhoods were summarily demolished and they were resettled far from their former domiciles.

The Huangpu greenways and parks are fully accessible and are widely used, especially on weekends. They are part of a plan to transform a "parks within the city" model to one more like "the city within parks." But affluent residents of the waterfront neighborhoods, like Qiantan, are the most immediate beneficiaries. They enjoy the parks as their own backyards on a daily basis, and their lifestyles are the ones most at home in these elegant green domains. The ubiquitous security staff who police these spaces have an uncomfortable air. They are there to stiffen the rules, liberally posted on signs, but their awkward presence, supplemented by the more clearly out-of-place rural migrant workers who make up the landscape maintenance crews, reveals the class differences between employees and the targeted users. Visitors who exhibit behavior that does not conform to middle-class recreational etiquette are frowned upon. Prohibition of the use of amplifiers, for example, guarantees a noise-free envi-

---

69 *Shanghai Manual: A Guide for Sustainable Urban Development in the 21st Century 2018*, Annual Report, UN-Habitat (2018), https://unhabitat.org.
70 Yifei Li and Xiaohua Zhong, "'For the People' Without 'By the People': People and Plans in Shanghai's Waterfront Development," *International Journal of Urban and Regional Research* 45, no. 5 (September 2021): 835–47.

ronment with no unauthorized sonic disorder.[71] The class-conscious messaging extends to signage about how to behave around the plant life, reaching far beyond the most obvious admonition to refrain from picking flowers. Here, the civilizing portion of shengtai wenming comes into play. Green governance over urban space like this delivers a lesson in class civility. Violations may bring down fines, or deductions from the social credit scores—a performance measure, often compiled through facial recognition technology, used by the state to assess an individual's or a company's trustworthiness. Not surprisingly, having more children increases your social credit.

As with waterfront development in former port cities, including New York, the change of land use is a key part of what has been called "green gentrification." When brownfields are revitalized as environmentally friendly spaces, their value to a new class of users and residents is often advertised by visible signs of repurposing. Thus, in keeping with the template of postindustrial gentrification elsewhere, the Huangpu regeneration plan preserves landmark industrial features such as docks, tower cranes, and observation towers as nostalgic reminders of an "authentic" prehistory.

A key part of the regeneration project involved the reinforcement and elevation of the waterfront walls lining the river, and it could therefore be counted as a climate change adaptation effort. These dykes are the city's first line of defense against waterborne climate impacts. The chronically polluted Huangpu (the last tributary of the Yangtze before it flows into the sea) runs through the city's heart. Its tides are periodically swelled by climate-driven storm surges or flooding upstream in Lake Taihu, the river's source. These surges increasingly now rise to levels of five meters or more. Because of the low-lying population zones of the Pearl and Yangtze River deltas, China now has about a quarter of the world's total populated land that will be below the rising sea level by the end of the century: up to fifty-seven million Chinese are at risk of being underwater; rural areas far inland are not immune. In 1998, excessive rainfall caused

---

71 Jingyi Zhu, "Public Space and its Publicness in People-Oriented Urban Regeneration: A Case Study of Shanghai," *Journal of Urban Affairs* (2023): 1–20.

floods in Jiangxi, Hubei, and Hunan that affected as many as one-fifth of China's population; more than 3,000 people died, fifteen million were made homeless, and fifteen million farmers lost their crops.

Urban centers on vulnerable coastlines are the most at risk, however, and these include the majority of the world's megacities. Like the East Coast of the US, China's coastal sea levels are rising faster than elsewhere. Shanghai, which sits thirteen feet on average above the mean tide, is particularly vulnerable because it is built on alluvial mud deposited by the Yangtze River over thousands of years, and it is in the pathway of tropical cyclones or typhoons. Just two months after the Communist takeover in 1949, a catastrophic typhoon, combined with high tides, breached the levees and placed most of the city under six feet of water, killing 1,600 people and rendering 200,000 homeless. On average, sea level has risen 1.6 millimeters annually for the past fifty years (eighty millimeters total or 0.26 feet), and Shanghai's marine monsoon climate and urban heat island effect means that its average temperature is rising twice as fast as the national average. In addition, like Jakarta, Bangkok, Ho Chi Minh City, Lagos, and many other overdeveloped port cities located on river deltas, Shanghai is sinking. With its dense concentration of high-rise buildings, the city's center has subsided by three meters since the late-nineteenth century, and so it is now decidedly concave with respect to its coastal perimeters, and sits only two meters above sea level.

The Huangpu flood plain is protected by levees, periodically reinforced and raised as part of the state's adaptation plan. All told, there are more than five hundred kilometers of seawalls around the metro area of thirty million, stretching across the Hangzhou Bay and encircling the islands of Chongming, Hengsha, and Changxing. Along with a proposed mechanical gate, and advanced warning technologies, they are part of a sophisticated flood control system that is being prepared to absorb 200-year events, though it remains to be seen how it will withstand the kind of deadly combination of typhoon plus high tides that inundated the city in 1949. Because of their blunt environmental impact, seawalls, which engineers overwhelmingly favor, are a contentious way of defending cities against waterborne climate impacts, yet they have become the technology of choice in high-profile projects like Jakarta's Great Garuda or Lagos's Eko Atlantic.

Many landscape architects prefer permeable zones or restored marshlands which absorb excess rainfall or tidal surges and then return the water after the danger is over. Alternatively, ponds, wetlands, green roofs, bioswales, greenways and rain gardens can capture the surplus water for reuse, or recharge aquifers.[72] Rather than reinforce hard edged walls around major population centers, this alternative is more fluid in character, allowing cities to coexist on a more dynamic basis with their wetland buffers, especially when they are part of river deltas. The tug of war between hard edge solutions and nature-positive approaches is playing out in coastal cities across the world. In my own backyard, New York City, the Army Corps of Engineers proposed building fifty miles of seawalls, levees, and berms between twelve and twenty feet tall, along some of the most scenic waterfront areas.[73] The coastal resiliency plan that was eventually approved for Lower Manhattan is much more integrated, combining hard infrastructure with parks and heavily landscaped, living shorelines.[74] In 2013, Chinese authorities jumped on the concept of the "sponge city," and Shanghai, with its formidable concentration of hard, impervious surfaces, was chosen as one of the thirty pilots. Select parks in each district were redesigned with sponginess in mind, and Lingnang, a new seventy-nine square-kilometer city at the southeastern tip of the metro area was specifically planned as an entire urban sponge.

One of the pre-2013 sponge prototypes, the prize-winning Houtan Park, on the eastern bank of the Huangpu, was presented as an ecological showpiece for the 2010 World Expo. Designed by Kongjian Yu, the landscape architect most associated with sponge cities, this linear thirty-five acre park, anchored by a mile-long constructed

72 Faiza Chikhi, Chuangcheng Li, Qunfeng Ji, and Xilin Zhou, "Review of Sponge City Implementation in China: Performance and Policy," *Water Science and Technology* 88, no. 10 (2023): 2499–520.

73 Robert Yaro and Daniel Gutman, "The Plan to Save New York From the Next Sandy Will Ruin the Waterfront. It Doesn't Have To," *The New York Times*, June 15, 2023, https://www.nytimes.com.

74 NYC Lower Manhattan Coastal Resiliency Project, https://www.nyc.gov.

wetland zone, was created on the former site of a waterfront steel factory. The wetlands, constructed around traditional riparian terraces, not only control flooding but also actively clean the polluted river water while slowing down its movement. Crops and bamboo are planted in the terraces, and reclaimed industrial structures and materials form parts of the pedestrian network through the park. Yu has insisted that his parks are "productive" insofar as their elements perform ecological services, mimicking the natural hydrologic cycles. Although the labor they perform is supposed to make them cost-effective, sponge parks like Houtan, let alone cities, are not cheap to construct. China has adopted sponge cities as a national policy, but the economic pathway from pilot projects to standardized applications will be steep.

While its design might have been replicated and scaled up, Houtan Park is only a small portion today of the forty- five kilometer Huangpu greenway that has been built out. Most of the greenway is composed of the impervious pavement that does nothing to absorb stormwater or control runoff, while the hardening of the flood control barriers themselves is a clear rendition of the old school hydrology, whose managers have built one large dam a day since 1949, immeasurably more than any other country. In some places, murals have been painted on the levee walls, depicting pastoral scenes with wild animals, and, occasionally, dolphins and other forms of marine life are portrayed pushing their snouts through "breaks" in the wall. The tableaux are a lovable illustration of coexistence with the natural environment, while the breaches suggest a more alarming future scenario.

TOMORROW'S WOMEN

After returning from Jiangxi, I spoke with two Shanghainese women who helped me clarify some outstanding questions. Xiaoyan Gao was the first of my interviewees who volunteered any opinions about climate change. She had majored in biology in college before pursuing an MBA at an American university. While there was nothing in either course of study about climate change, she told me that she applied what she learned, respectively, about the life sciences and business to form her own opinions about the impact of CO_2 emissions. "I don't think we are doing enough to guarantee the future of life on this planet," she affirmed, "at least not enough for me to justify bringing children into this world." Gao said that she tried to live frugally and maintain a small carbon footprint. "I couldn't do that if I started a family," she conceded, and this was not only a concern about herself. "I worry anyway about whether any of our children could have a good life in a bad climate."

Gao identified herself as a CPC member, but did not hesitate to take its leadership to task. When I asked about Xi's shengtai wenming, she said, "it could be a beautiful reality, if we really meant to do what it says, but it's like an opera about the GDP—and just another way for officials to stay in their positions." What would her alternative pathway look like, I asked? "The socialism I grew up with can still make a big difference in saving the planet," she mused, "but only if it listens to women like me, because we are the ones who can see more clearly what needs to be done." Gao, who worked for an NGO, had been a first-generation college student, and she credited her views, including her feminism, to her education, both in China and abroad. She acknowledged the privilege of women in her position, but did not regard that as a reason to hold her tongue. "You can't give us all of this knowledge," she observed, "and then expect us not to use our voices."

This was not the way that Beijing saw things. Women had lost a lot of ground under Xi's rule, with fewer now serving in roles of power, both at the national and provincial levels, and none at all in the Politburo. His administration cracked down on a resurgent feminist movement as part of its repression of civil society and has pushed LGBTQ+ advocates to the margins. In that context, it is difficult not to see the government's pressure on women to reproduce as anything but a rearguard effort to restore patriarchal norms and safeguard GDP growth. Gao did not disagree with this interpretation even though her choice to remain childless was guided primarily by her own ecological views. "My loyalties are divided," she coolly reminded me.

She recommended that I speak with Ningli Zhang, a college classmate, and we met the next day at the café inside my hotel after she finished work at a kindergarten in Pudong. "Many of the preschools have closed down," she reported, "because of the lack of demand, or they are now operating for seniors." Like most of the parents who used the facility, Zhang had only one child, a five-year-old son, and she was going through a divorce (in 2021, the government introduced a cooling-off period for divorce proceedings in order to boost birth rates). Even before she and her husband separated, she said she had no intention of adding a sibling. "There are enough Chinese in the world," she joked, "haven't you been in the Shanghai metro at rush hour?" "It would be better," she added, in a more serious vein, "to have more trains, and less people with cars."

Like her old friend Xiaoyan, she was careful with her finances, but in her case, her thrift was accompanied by a visceral dislike of the consumerism that had captured China in recent years. The retail boom was even more pronounced in Shanghai, with a longstanding reputation as a shopper's mecca. "I get dizzy and sick in a shopping mall," she reported, "and have to run out of there fast." Zhang recognized that her country had risen from poverty. "We have worked hard to have more, but not this kind of more," she said, gesturing around her, as if to rows of shiny merchandise—"these things that people are obsessed with buying are just like the opium that the British got us addicted to." The ads are directed at women, she noted, "and they are often pictured in very happy family units." Zhang continued, "we did not go to university to become compulsive shoppers," concluding that "you don't have to be a feminist, as I am, to see that this is wrong."

Zhang also told me she was a social Christian, "not a Marxist," and so that meant she believed "not in revolution, but in the evolution of society towards a better life." Evangelists were behind the extraordinary growth of Christianity in China in recent decades, but she was not one of those, she assured me. "My faith comes from what we can build together—for example, more schools, hospitals, theaters, parks, churches—not from my own hope, as an individual, to be filled with God or divinity." Her father was an engineer who had helped to design bridges, and she had inherited some of his enthusiasm for improving public infrastructure. Even though it was overcrowded, the Shanghai metro, she pointed out with visible pride, "worked very well, all the time," as did "all of these solar and wind farms that bring us such good energy." Her trust in technology was almost contagious.

Shanghai had more than its share of middle-class, female individualists, eager to deploy their newly available liberty to make independent choices. I interviewed quite a few of them.[75] Gao and Zhang seemed to be cut from a different cloth; principled, credible, and confident about the power of cooperation to usher in the future. They did not fit the *quality* profile of human capital so desired by the CPC's new generation of model workers, but neither had they turned their back on their socialist heritage. Rather than expressing cynicism or dread about what lay ahead, they had convictions about the role they could play in the country's ability to parlay its long record of achievements into a more authentic version of shengtai wenming. When we finished talking, I had the sense that I had met some of "tomorrow's women."

Gao's sense of responsibility in the face of the climate crisis was framed by women's empowerment but it was not confined to her control over reproductive choices. She thought that the party to which she belonged, at least nominally, had an obligation to forge an eco-socialist path. Zhang shared the technocratic outlook of the CPC leadership, mostly engineers by tradition, but her rejection of market capitalism was at odds with their economic policies. As for the fraught relationship between population and climate, she seemed to be more concerned about the high-carbon lifestyles of her urban peers than the number of people in the world.

No doubt, I warmed to their views because they reflected many of my own. But how widely shared were these opinions; and what were the prospects that they might take a more organized political form among like-minded Chinese women and their allies? Those questions were beyond the scope of my investigation. As I left Shanghai, I took away the hope that there were many more women like them, and that they would find each other, and make history under conditions not of their choosing.

---

75 See Jiaran Zheng, *New Feminism in China: Young Middle-Class Chinese Women in Shanghai* (Singapore: Springer, 2016).

Spring in Houtan Park, Shanghai, April 2024.

CONCLUSION
REPRODUCING THE FUTURE

Five months after my Shanghai sojourn, more than 400,000 people were evacuated in advance of Typhoon Bebinca, by far the strongest storm to hit the city in seventy-five years. Luckily the typhoon did not make landfall during a high tide, as it had done in 1949, and so the damage was not as extensive. It was by no means an outlier in the storm season of 2024; among the many extremes reported, the Philippines were battered by no less than six major typhoons over a thirty-day period, while the Atlantic season saw a record eleven hurricanes, leaving behind a widespread trail of infrastructure devastation, casualties, and massive economic losses. The super-destructive "storms of my grandchildren," to quote the title of James Hansen's 2009 book, were appearing sooner than even its author (the most outspoken of climatologists) had predicted.

In February 2025, Hansen and his colleagues published research about the impact of Arctic ice melt on prevailing ocean currents, arguing that "a shutdown of the Atlantic Meridional Overturning

Circulation (AMOC) is likely within the next twenty to thirty years,"
barring drastic actions to reduce global warming.[1] This event would
be a "point of no return," resulting in several meters of sea rise, rapid
swings in temperature, and a spike in extreme climate-driven disasters. Their warning rebutted the more conservative IPCCC claims
that the AMOC's system of vital currents (including the Gulf Stream),
which conveys warm water from the Southern Ocean near Antarctica
up to the Arctic Sea, would only slow, but not shut, down over the
next century.

By the time of the article's publication, that timeline of twenty
to thirty years might have been further shortened by a different kind
of shutdown. The second Trump administration had begun to ruthlessly purge data and dismantle environmental programs with a view
to withdrawing the US from all forms of climate action and research.
The evaporation of official US support for decarbonization initiatives
for at least the next four years means that advocates of state-level
responses to the climate emergency will need to look elsewhere.
China's combination of strong state authority, technological resolve,
and mass labor power make it the only likely contender to bring
about significant climate action within the ever-narrowing timeframe
before a more catastrophic Anthropocene arrives. Only China has the
wherewithal to manufacture a low-carbon energy economy at appropriate scale, or "build the arks" to survive the coming floodwaters, as
depicted in the Hollywood disaster film *2012* (a long-running #1 at the
box office in China) that I mentioned in this book's Introduction. Its
leaders seem to relish the praise for showing green leadership on the
world stage, and its proven capacity to implement weather modification efforts and construct mega-infrastructural projects also makes it
the most viable candidate to deploy the fraught, salvation option of
geoengineering.

For all of its formidable headway in producing clean energy, the
CPC's commitment to maintain high levels of GDP-driven growth
has also made it a voracious extractor and consumer of coal, and so

---

1   James Hansen et al., "Global Warming Has Accelerated: Are the United
Nations and the Public Well-Informed?," *Environment: Science and Policy for Sustainable Development* 67, no. 1 (February 2025): 6–44.

its carbon balance sheet is not all that different from other leading fossil producers, who flaunt their heroic domestic achievements in lowering emissions while ramping up hydrocarbon production for export to poorer countries that cannot afford renewable technologies. In addition, China has little in the way of civil society and therefore its current pathway to a greener future is fundamentally nondemocratic. Government officials closely monitor the activities of all non-state actors, and they have little tolerance for citizen-led initiatives. Commons-based alternatives to state capitalism, such as cooperative or solidarity economies, are all but nonexistent. In other countries, where communities befouled by corporate profiteers can mobilize members to fight for more public regulation or ownership, their Chinese counterparts are limited in their leeway for self-organization, let alone the ability to leverage change from below. That top-down habit of ruling is why the fading inclination to procreate has become so contentious; birth rates are considered to be an affair of the state, but their increase is beyond its will.

Nevertheless, the Chinese government's anxiety about population decline highlights the neglect of social reproduction in the larger debates about how to implement climate action. For several decades, forecasters of ecological collapse obsessively pinpointed "overbreeding" as the key cause. More recently, CO_2 has largely taken its place as the threatening object that needs to be reduced in quantity. Populationism has not dissipated, though it is more and more the preserve of the resurgent proto-fascist right, mobilized against the flow of immigrants, and in defense of white ethnic "blood and soil." Propagating their own kind, in the face of the "civilizational suicide" of US fertility decline, has become an obsession of white elite figures like Elon Musk, who has fathered fourteen children and allegedly wants to offer his sperm freely in order to people a Mars colony.[2] But the overwhelming attention to combatting carbon emissions among progressives has sidelined many longstanding environmental concerns and challenges, including a more honest reckoning with the role of reproduction in any just transition to a green future.

---

2   Kirsten Grind, "Elon Musk's Plan to Put a Million Earthlings on Mars in 20 Years," *The New York Times,* July 11, 2024, https://www.nytimes.com.

This negligence extends to advocates of swift and just action on CO_2, who are themselves divided on the methods and philosophy to guide us there.[3] Supporters of degrowth, for example, who urge an overall shrinkage of expansive economic activity, are at odds with champions of the Green New Deal who maintain that "green growth" is the most painless path to climate stabilization. The former want to promote popular, self-organized alternatives to GDP-driven development and extraction; most degrowth scenarios entail reordering capitalist economies in favor of a radical redistribution of income and wealth. Green New Dealers have pinned their hopes on securing large-scale, Keynesian investments in clean energy to ward off the worst CO_2 impacts; most versions of a Green New Deal look to maintain or raise standards of living by modernizing economic production through decarbonization.[4]

There are many takeaways from the debates between these two left camps. Degrowthers point out that a single-minded focus on making a great leap forward in clean energy production overlooks the problems of uneven development and climate injustice, and that green growth leaves the ruinous capitalist system of commodification and labor exploitation largely intact.[5] So far, the gains in renewable

---

3   See Stefania Barca, *Workers of the Earth: Labour, Ecology and Reproduction in the Age of Climate Change* (London: Pluto Press, 2024); Mary Mellor, "An Eco-Feminist Proposal," *New Left Review* 116/117 (March–June 2019); Silvia Federici, *Revolution at Point Zero: Housework, Reproduction, and Feminist Struggle* (Oakland: PM Press, 2020).

4   Benjamin Kunkel and Lola Seaton (eds.), *Who Will Build the Ark?* (London: Verso, 2023) collects some of the debates between degrowthers and Green New Dealers that appeared in the pages of the *New Left Review*.

5   Among the many books on degrowth are: Jason Hickel, *Less is More: How Degrowth Will Save the World* (London: William Heinemann, 2020); Matthias Schmelzer, Aaron Vansintjan and Andrea Vetter, *The Future is Degrowth: A Guide to a World Beyond Capitalism* (New York: Verso, 2022); Nathan Barlow et al. (eds.), *Degrowth and Strategy: How to Bring About Social-Ecological Transformation* (London: Mayfly, 2022); Serge Latouche, *Farewell to Growth* (Cambridge: Polity, 2009); Giorgos Kallis, Susan Paulson, Giacomo D'Alisa, and Federico Demaria, *The Case for Degrowth* (Cambridge: Polity Press, 2020); Giacomo D'Alisa, Federico Demaria, and Giorgos Kallis (eds.), *Degrowth: A Vocabulary for a New Era* (Abingdon: Routledge, 2014); Iris Borowy and Matthias Schmelzer (eds.), *Degrowth: A Vocabulary for a New Era* (Abingdon: Routledge, 2014); Iris Borowy

energy have resulted in a supplement to, and not a replacement of, fossil fuel extraction. The latter continues to expand while the ever cheaper prices of solar and wind power are dampening the animal spirits of investors. So too, capitalism's deep-seated propensity to destroy the freely available and highly profitable inputs of nature—air, water, minerals, fossils, topsoil, the atmosphere—means that it can never be trusted, least of all when clothed in green, to provide solutions to ecological catastrophes of its own making. On the other hand, while we might revel in the prospect of imposing "de-growing pains" on the beneficiaries of ultra-luxury consumption and carbon-intensive waste, there is no popular appetite for programs to curtail growth if they require hardship and austerity for the less affluent. As such, political feasibility should not be sidelined as a factor when drawing up green industrial policy.[6]

Despite differences between the two camps, there are overlaps and points of agreement, with neither finding any comfort in the current US policy environment of renewed backing for fossil-fueled

---

and Matthias Schmelzer, *History of the Future of Economic Growth: Historical Roots of Current Debates on Sustainable Degrowth* (Abingdon: Routledge, 2017); Tim Jackson, *Prosperity Without Growth: Economics for a Finite Planet* (London: Earthscan, 2009); Kohei Saito, *Slow Down: The Degrowth Manifesto*, trans. Brian Bergstrom (New York: Astra House, 2024); Timothée Parrique, *Ralentir ou périr: L'économie de la décroissance* (Paris: Seuil, 2022).

6   Among the books about the Green New Deal are: Ann Pettifor, *The Case for the Green New Deal* (London: Verso, 2019); Noam Chomsky and Robert Pollin, *Climate Crisis and the Global Green New Deal: The Political Economy of Saving the Planet* (New York: Verso, 2020); Kate Aronoff, Alyssa Battistoni, Daniel Aldana Cohen, and Thea Riofrancos, *A Planet to Win: Why We Need a Green New Deal* (New York: Verso, 2019); Guido Girgenti and Varshini Prakash (eds.), *Winning the Green New Deal: Why We Must, How We Can* (New York: Simon & Schuster, 2019); Naomi Klein, *On Fire: The Burning Case for a Green New Deal* (New York: Simon & Schuster, 2019); Jeremy Rifkin, *The Green New Deal: Why the Fossil Fuel Civilization Will Collapse by 2028, and the Bold Economic Plan to Save Life on Earth* (London: St. Martin's Press, 2019); Jeremy Brecher, *The Green New Deal from Below: How Ordinary People Are Building a Just and Climate-Safe Economy* (Urbana: University of Illinois Press, 2024); and with a significantly decolonial orientation, Max Ajl, *A People's Green New Deal* (London: Pluto, 2021) and the Red Nation, *The Red Deal: Indigenous Action To Save Our Earth* (Philadelphia: Common Notions, 2021).

growth. Like it or not, catastrophic harm to life, land, and the built environment is inevitable in the near future, and so degrowth, even if measured in terms of GDP loss, will come anyway at the hands of a rapidly changing climate. Disaster capitalism will feed off such dire circumstances; the wealthiest will be well insulated, and the more vulnerable are liable to be sacrificed through triage management. Surely, it would be better to voluntarily adopt a post-growth program than have one imposed by extreme global warming. If another world is possible, it will have to be intentionally planned in anticipation of losses that are baked in as a result of atmospheric CO_2 build-up. That is not defeatist, nor anti-utopian: it is the prerogative, if not the obligation, of people who have choices left to make.

When I wrote *Strange Weather: Culture, Science and Technology in the Age of Limits*, my 1991 book about global warming and environmental limits, it was greatly influenced by Murray Bookchin's thinking about social ecology. For Bookchin, who was the first major left thinker to link capitalist growth with global warming, the roots of the ecological crisis and the domination of nature lay in human society, i.e., in the domination of one group over another. When he spoke about a post-scarcity society, he envisaged one in which communities made choices of their own accord, and not in response to some manufactured state of artificial scarcity.[7]

The bogus scarcity trick Bookchin warned us about is alive and well today. The examples offered in this book include the colonial strategy of water apartheid in Palestine and the "resource-saving" campaign of immigrant exclusion in Arizona. Nor can we afford to disregard Bookchin's healthy distrust in state authority. Just consider that around 75 percent of the world's hydrocarbon supplies are in the hands of state-owned companies like Saudi Aramco and ADNOC (Abu Dhabi National Oil Company), the antiheroes of this book's report from COP28. Every well, pipeline, and terminal being constructed and operated by these national firms has the political backing of the states in question.[8] Going on the evidence of the expansion of

---

7   Murray Bookchin, *Remaking Society* (Montreal: Black Rose Books, 1992).

8   Because energy, unlike other trade commodities, is a matter of national security, even the private fossil companies operate with the regulatory endorsement of territorial governments.

fossil fuel production since 2022, none of them are interested in keeping the oil in the soil. That is not to say that public ownership in the energy sector is a lost cause. Far from it. Public utilities, for example, produce cheaper, safer, and cleaner electricity, and are usually more reliable than the privately-owned ones. Indeed, a timely transition to clean energy is unconceivable without the exercise of large-scale public power, and ultimately, state-level efforts to upgrade national and international grids. However strongly they resonate with Bookchin's libertarian municipalist ideals, our local, small-scale energy commons will not be sufficient to do the job.

That is why the public-commons model is the most promising pathway for a sustainable future. Following this template, grassroots social justice or movement organizations forge alliances with the officials who oversee public institutions with the goal of self-managing their public goods and services. The outcome is an antidote to the neoliberal model, where officials contract out services to the private sector in the name of efficiency and with little or no regard for any "just transition."

International action for climate justice is more challenging, to say the least. The UNFCCC process has been the most ambitious effort at interstate cooperation since the UN was formed in 1948 to keep the postwar peace. Nevertheless, it has failed miserably to produce an effective "change agent"; in Alyssa Battistoni's words, the UNFCCC has become "not so much an emergent global sovereign as a charity raiser" for poor nations.[9] In the meantime, the masters of capital make their own rules in other forums where they don't have to perform on the stage of climate philanthropy at UNFCCC meetings. So far, the task of building an effective opposition to fossil power—a truly international alliance of anticapitalist movements— has proved elusive.

In the 1990s, La Via Campesina emerged as a Global South coalition that comprises "millions of peasants, landless workers, indigenous people, pastoralists, fishers, migrant farmworkers, small and medium-size farmers, rural women, and peasant youth." Their voice, reflecting the perspective of the "environmentalism of the poor," was

---

9 Kunkel and Seaton, *Who Will Build the Ark?*, 223.

well represented in the landmark *People's Agreement on Climate Change and the Rights of Mother Earth* forged in 2010 in Cochabamba.[10] But where do urban workers in fully industrialized societies find points of identification in the Cochabamba ethos of *buen vivir* [living well], or living in harmony with nature? For example, the labor movement in the Global North has long been hitched to the consumer-driven wagon of *living better*, a principle that stands in diametrical opposition to buen vivir because it is associated, in the Cochabamba mind, with resource plunder, materialist growth, and climate debt. While they may share the same enemies, enriched and empowered by forty years of neoliberalism, the priorities of Northern urban workers as promoted by the Green New Deal are not necessarily the same as those of *campesinos* faced with dispossession and displacement from their land. Everyone's lives may be better sustained by commoning practices such as agro-ecology, a universal basic income, or energy sovereignty, but much remains to be done to bridge the divide between these two populations.

During the hiatus of the COVID pandemic, large portions of the world's population withdrew from routine, workplace productivity, emissions plummeted, and the planet got a welcome respite. The return of wildlife to uncongested cities and the clearing of polluted skies were breathtaking reminders to people—told for so long that "there is no alternative"—that another world was possible after all. The lockdown occasioned outpourings of mutual aid, massive government disbursements of basic income, and creative use of time, and it was followed by a widespread disinclination, termed the "Great Refusal," to return to ungratifying workplaces in order to staff the machineries of profit. Yet the pandemic economy's reliance on "essential workers" highlighted how indispensable care work is to the day-to-day operations of our lives. Long taken for granted and dominated by women and/or workers of color, these occupational

---

10 See Ashley Dawson's useful survey of the Cochabamba legacy, in *Environmentalism from Below: How Global People's Movements Are Leading the Fight for Our Planet* (Chicago: Haymarket, 2024); and Juan Martínez Alier, *The Environmentalism of the Poor: A Study of Ecological Conflicts and Valuation* (New Delhi: Oxford University Press, 2005).

roles were thrust into the limelight. With the growth-driven engines of capitalism at a standstill, the care economy was revealed as the only one that really mattered, and the centrality of its workers pointed the way forward to a more sustainable social ecology.

During the lockdown, care work was not only frontline, and officially essential, but its net of informal participants expanded, as did the range of work falling under its rubric. Confined to their homes, more and more of the general population contributed to the tasks of cooking, feeding, cleaning, educating, tending, consoling, entertaining, maintaining, and repairing. Over time, it became clear (especially to men not usually called upon) how central these functions and responsibilities are to a truly sustainable economy. Care work is not a secondary, supportive sector. To all intents and purposes, it is the primary one. If planetary life is to survive, care needs to be elevated to a top priority of our societies, for the long term and not simply in times of public health emergencies or climate-driven disasters. This essential work, whether in childcare, nursing, midwifery, teaching, community health, domestic labor, or care for the elderly, disabled, and unhoused can no longer be a marginal sector of undercompensated or unwaged labor, it will have to be undertaken by everyone. Part of that common enterprise will involve bringing new meaning to the old call for the dignity of labor. As a shared obligation, the work of care extends to protection not only of the land, rivers, forests, and biodiversity, but also of communities most impacted by the changing climate. Environmental justice is nothing if not an undertaking of care and accountability.

In countries with graying populations, the growing burden of care is all too obvious, as we saw during our visit to Stella's Jiangxi orphanage, now peopled with retirees who may be the last legatees of the Maoist "iron rice bowl" of cradle-to-grave security. But the need is just as great at the other end of the life cycle. Decades of neoliberal divestment in welfare services and the stripping of rights and protections have weakened the ability of women to give birth by choice and with any expectation of security. From the perspective of social reproduction—how a society sustains its general population—these assaults on the provision of basic social goods and the privatization of related public assets are apiece with the deregulation of environmental protection.[11] That such aggressive measures are cheered on by the creditor and investor classes—the true "super-predators"—is further evidence of how capitalism destroys its own conditions of reproduction.

The resulting threats to women's reproductive health and freedoms jeopardize the resources that we need for resilience in the face of the climate crisis. That is the real story about population growth, not the one about overbreeding on the part of poor women in the Global South. After all, without a healthy pathway to producing future generations, climate change is guaranteed to turn into an "extinction-level event" for humans, as it already is for millions of other species. To ensure the planet remains habitable, we need a resourceful and imaginative next generation, raised in security, and armed with endurance skills and a cooperative mindset.

The ethical dilemma of whether to bring children into a world impaired by global warming has consumed a lot of attention and no end of books and magazine articles.[12] Eco-anxiety has even been diagnosed as a factor in the baby bust's decline in fertility.[13] A wide range of voices, from moral philosophers to grassroots groups like BirthStrike for Climate, now promote non-procreation as a righteous

---

11 See Giovanna Di Chiro, "Living Environmentalisms: Coalition Politics, Social Reproduction, and Environmental Justice," *Environmental Science, Political Science* 17, no. 2 (April 1, 2008): 276–98.

12 Some books probing the dilemma are Elizabeth Rush, *The Quickening: Antarctica, Motherhood, and Cultivating Hope in a Warming World* (Minneapolis: Milkweed Editions, 2023); Emily Raboteau, *Lessons for Survival: Mothering Against the Apocalypse* (New York: Holt, 2024); Meghan Elizabeth Kallman and Josephine Ferorelli (eds.), *The Conceivable Future: Planning Families and Taking Action in the Age of Climate Change* (London: Rowman & Littlefield, 2024); Gina Rushton, *The Parenthood Dilemma: Procreation in the Age of Uncertainty* (New York: Astra, 2023); Rachel Wiseman and Anastasia Berg, *What Are Children For? On Ambivalence and Choice* (New York: MacMillan 2024), and Mara van der Lugt, *Begetting: What Does It Mean to Create a Child?* (Princeton: Princeton University Press, 2024).

13 Jade Sasser analyzes the racial differences that affect our "eco-emotions" in *Climate Anxiety and the Kid Question: Deciding Whether to Have Children in an Uncertain Future* (Oakland: University of California Press, 2024).

cause.[14] This kind of pressure on women is misplaced; it is only the latest, albeit the most genteel, version of the populationists' identification of excess reproduction as the root source of ecological collapse. Let us agree that the real villains sit in the boardrooms of corporations, especially fossil extractors, and that we can rightly point our finger at their stranglehold on the apparatus of democratic power.

Having contributed to the work of parenting, teaching, and caring for the well-being of young people in a world that is burning up, what can I say? The act of bearing a child should not be dismissed as an anti-green choice, if only because we have no inkling of how our offspring will act or what they will achieve in their lives. Nor should childrearing be unreservedly celebrated as a virtuous expression of hope in a time when most of the indicators are pointing in another direction.[15] We don't need macro-economists to warn us about the doom loop through which one negative condition—fertility collapse—leads to another—financial stagnation. Instead, we can simply use our common sense to see that the presence of children obliges a community to prepare for the future. And, that the pursuit of justice is impossible without a vision, if not a blueprint, of a better world for them to inherit.

The best place to learn that lesson is in the Gaza Strip, where children made up almost half of the population before October 2023, and where their death (15,600 by March 2025) or that of their parents has been arguably the most barbaric feature of the genocide. With a fertility rate of 3.38, Gaza has long defied the demographer's rule that a population with high levels of education, especially for women, will show a declining birth rate. Given the Palestinian history of displacement, impoverishment, and premature death at the hands of military attacks and blockades, it is no surprise that reproducing

---

14 Among the more extreme perspectives are David Benatar, *Better to Have Never Been: The Harm of Coming Into Existence* (Oxford: Oxford University Press, 2006) and Lee Edelman, *No Future: Queer Theory and the Death Drive* (Durham: Duke University Press, 2004).

15 As Kafka wryly commented to his biographer Max Brod, "there is plenty of hope, infinite hope—just not for us."

a numerous next generation would be a top priority, or that their education (in a society with one of the world's highest literary rates) would take precedence. For a hundred years or more, the bloodlines of Palestinian families have been at risk of erasure. Under these circumstances, bearing and rearing children have long been seen as acts of resistance in themselves—as blows against the Occupation and Israel's efforts to win demographic dominance in the lands between the river and the sea.

As the genocide took hold, the desperate efforts of Palestinians to care for babies and infants, or to salvage youth, took on a heightened dimension. Israel's targeted attacks on Gazan access to health, medical care, and education decimated the system of social reproduction in the Strip, and were widely seen as a systematic endeavor to put to a stop to the next generation of Palestinians.[16] How would the children who had not been maimed or orphaned overcome the cumulative mental harm from their experience of what psychiatrists call "complex continuous trauma" (CCT) since October 2023? And how would Gazan families, menaced by Trump's and Netanyahu's new threats of permanent eviction and displacement, find the wherewithal to survive, propagate, and nurture their offspring?

The genocide in Gaza was a sharp escalation of existing patterns of harm and violence. A 2020 study of Palestinian children in the Gaza Strip found that 88 percent had experienced personal trauma, 84 percent had witnessed trauma in others, 88 percent had observed property demolition, and that as many as 54 percent suffered from post-traumatic stress disorder (PTSD).[17] Children born and bred in such a war-torn environment, where mental and physical harm is omnipresent, are not necessarily more resilient.[18] But when they grow up in a Palestinian culture anchored by the principle of *sumūd* [steadfastness], and the refusal to be defined by the trauma and vio-

---

16 Palestinian Centre for Human Rights, *Generation Wiped Out: Gaza's Children in the Crosshairs of Genocide*, December 31, 2024, https://pchrgaza.org.

17 Basil El-Khodary et al., "Traumatic Events and PTSD Among Palestinian Children and Adolescents: The Effect of Demographic and Socioeconomic Factors," *Front Psychiatry* 11, March 30, 2020.

18 See Yara M Asi, *How War Kills: The Overlooked Threats to Our Health* (Baltimore: Johns Hopkins University Press, 2024).

lence, they are raised to see their world as life-affirming even when it is bent into the most asphyxiating forms by the colonizer.[19] Their upbringing reflects the common will of a whole society to thrive, and to expect a future where others might not.

In February 2025, when Donald Trump floated his plan to evacuate Gazans to make way for a real estate development, Palestinians (amplified by their worldwide allies) affirmed the most literal meaning of sumūd—the resolve to remain on the land, no matter how ravaged or contaminated. It was an act of will that resonates with refugees everywhere, and especially those whose homes, welfare, and livelihoods are under threat from climate change. This is how Naomi Klein described it, in making the case for *climate sumūd* in her 2016 Edward Said lecture, "Let Them Drown: The Violence of Othering in a Warming World":

> [Sumūd is] a word most associated with places like Hebron and Gaza, but it could be applied equally today to residents of coastal Louisiana who have raised their homes up on stilts so that they don't have to evacuate, or to Pacific Islanders whose slogan is 'We are not drowning. We are fighting.' In countries like the Marshall Islands and Fiji and Tuvalu, they know that so much sea-level rise is inevitable that their countries likely have no future. But they refuse just to concern themselves with the logistics of relocation, and wouldn't even if there were safer countries willing to open their borders—a very big if, since climate refugees aren't currently recognised under international law. Instead they are actively resisting: blockading Australian coal ships with traditional outrigger canoes, disrupting international climate negotiations with their inconvenient presence, demanding far more aggressive climate action.[20]

Almost a decade after her comments, climate justice warriors today have even more reason to see how the struggle for Palestinian lib-

---

19 Lara Sheehi and Stephen Sheehi analyze this dynamic in *Psychoanalysis Under Occupation Practicing Resistance in Palestine* (New York: Routledge, 2021).

20 Naomi Klein, "Let Them Drown: The Violence of Othering in a Warming World," *London Review of Books*, 38, 11, June 2016, https://www.lrb.co.uk.

eration holds a key to conceiving the future we want, even when it
seems least feasible.

For a literal illustration of conception under duress we can look
to the example of Walid Daqqa, the much-admired Palestinian pris-
oner who, after being repeatedly denied conjugal visits, smuggled
sperm out of his Israeli prison so that his spouse Sana Salama could
give birth to their daughter Milad in 2020. He described his child as
"my message to the future."[21] Through his writings and his model of
unyielding, Daqqa delivered many other lessons about the art of per-
sisting. Incarcerated in 1986, he was martyred by his jailers' medical
negligence on April 7, 2024, having inspired countless fellow prisoners and others beyond the walls.

Milad's birth, however momentous, was not a lone example.
Dozens of Palestinian children have been conceived from sperm-
smuggling since the practice was first documented in 2012. It has
been acclaimed as an "affirmation of life" against a carceral order
aimed at halting the intergenerational legacy of Palestinian leaders and annihilating its underlying mentality.[22] In his well-known
text "Consciousness Molded or the Re-Identification of Torture,"
Daqqa analyzed the Israeli jailers' efforts not just to "break" prisoners
through torture and deprivation, but to rewire their political sensi-
bilities into more dissociated forms.[23] More than anyone, he documented how the practice of sumūd was embodied in the individual
and collective discipline of incarcerated Palestinians. As he put it, in
a 2005 letter that reflected on the "parallel time" experienced by those
who are held in prisons:

> To stop feeling shocked and bewildered, to stop feeling people's
> sorrows (any people), to be numb in the face of atrocity (any
> atrocity)—these became my daily nemesis, how I measured
> my sumūd and steadfastness. A person's mental essence is their

---

21 See Basil Farraj, "In Memory of Walid Daqqah," *Peoples Dispatch*, April 13,
2024, https://peoplesdispatch.org.
22 Layal Ftouni, "'They Make Death, and I'm the Labor of Life': Palestinian
Prisoners' Sperm Smuggling as an Affirmation of Life," *Critical Times* 7, no. 1
(2024): 94–109.
23 Walid Daqqa, *Consciousness Molded or the Re-identification of Torture* (Beirut,
Arab Scientific Publishers, 2010).

will, their physical essence is their labor, and their spiritual essence is their ability to feel. Being able to feel for people, feeling the pain of humanity, is the essence of civilization.[24]

How can we honor the lessons of Walid Daqqa, and those of influential female counterparts like Khalida Jarrar, who has called attention to the Israeli incarceration of women and children? Jarrar told us, memorably, that "hope in prison is like a flower that grows out of stone."[25] How should we acknowledge and amplify the enduring resistance of myriad other Palestinians living under the threat of eradication? For many of us, their ordeal and their steadfast fight for liberation from generations of captivity and displacement has come to personify the challenge we confront in the face of the much larger threat to planetary life itself. This book has tried to document the threads that connect these two orders of threat as they weave together the regimes of fossil fuel fundamentalism, ethnic nationalism, population control, resource hoarding, and capitalist dominion. There are no master blueprints for those actively working toward freedom in the years ahead, but since climate justice is quite literally an existential struggle, our efforts will surely be enlivened by the spirit of the Palestinian tenet that "existence is resistance."

---

24 Daqqa's letter, translated by Julia Choucair Vizoso, is in "Victory Over My Jailer": The Afterlife of Revolutionary Walid Daqqa's Steadfast Love," *Public Source*, April 17, 2024, https://thepublicsource.org.
25 Palestinian Feminist Collective, "Free Khalida Jarrar, End the Global Prison-Industrial Genocide," September 2, 2024, https://palestinianfeminist collective.org.

ACKNOWLEDGMENTS

Reporting of this kind depends on the goodwill and assistance of many people. Among those I would like to thank are:

In Palestine—Amin Husain, Samer Karajeh, Amy Weng, Amar Husain, Jenna Sharkawy, Joyce Matos, and Nadine Fattelah, who co-authored parts of the "Resistance Farming" section of Chapter 1.

In Dubai—James Lynch from FairSquare, Joey Shea from Human Rights Watch, and Simeon Kerr from the *Financial Times.*

In Arizona—Stacey Champion, Sarah Roberts, Moyheddin Abdulaziz, Judy Bourg, Gail Kocourek, and Todd Miller.

In China—Vivian Du, Lili Lin, Diego Gullotta, Chris Connery, and Rebecca Karl.

Thanks are due to my daughters—Stella Ross-Gray, who motivated our visit to Jiangxi, and Zola Ross-Gray, who provided research assistance. The whole book is informed by the companionship and advice of Maggie Gray, my partner.

Parts of the book were previously published in the *Boston Review, Dissent,* and *The New York Review of Books,* where I benefited from the respective editorial skills of Deborah Chasman, Nick Serpe, Ratik Asokan, and Max Nelson.

At Common Notions, Malav Kanuga was a wise, enthusiastic publisher, and Erika Biddle did a scrupulous editing job on the manuscript.

The names of several people who agreed to speak with me have been changed to protect their identities.

ABOUT THE AUTHOR

Andrew Ross is Professor of Social and Cultural Analysis at NYU. A contributor to *The Guardian, The New York Times, The Nation,* and *Al Jazeera,* he is the author or editor of more than twenty-five books, including *Cars and Jails: Freedom Dreams, Debt, and Carcerality; Sunbelt Blues: The Failure of American Housing; Bird On Fire: Lessons from the World's Least Sustainable City; Stone Men: The Palestinians Who Built Israel; The Celebration Chronicles: Life, Liberty and Property Values in Disney's New Town;* and *Fast Boat to China: Lessons from Shanghai.* His most recent book is *Abolition Labor: The Fight to End Prison Slavery.* He is also a co-founder of several movement groups, including the Debt Collective, Gulf Labor Artists Coalition, and Decolonize This Place, and currently serves as secretary of the Faculty and Staff for Justice in Palestine national network.

ABOUT COMMON NOTIONS

Common Notions is a publishing house and programming platform that fosters new formulations of living autonomy. We aim to circulate timely reflections, clear critiques, and inspiring strategies that amplify movements for social justice.

Our publications trace a constellation of critical and visionary meditations on the organization of freedom. By any media necessary, we seek to nourish the imagination and generalize common notions about the creation of other worlds beyond state and capital. Inspired by various traditions of autonomism and liberation—in the US and internationally, historical and emerging from contemporary movements—our publications provide resources for a collective reading of struggles past, present, and to come.

Common Notions regularly collaborates with political collectives, militant authors, radical presses, and maverick designers around the world. Our political and aesthetic pursuits are dreamed and realized with Antumbra Designs.

www.commonnotions.org
info@commonnotions.org

COMMON
NOTIONS

MORE FROM COMMON NOTIONS

DECOLONIZE CONSERVATION
GLOBAL VOICES FOR INDIGENOUS SELF-DETERMINATION, LAND.
AND WORLD IN COMMON

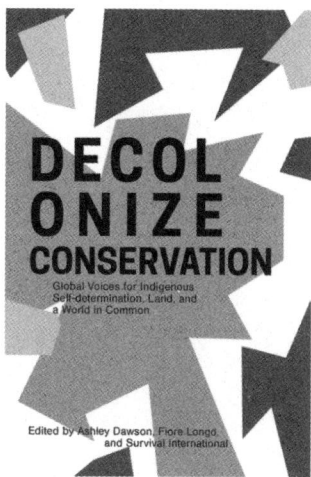

Edited by Ashley Dawson, Fiore Longo,
and Survival International

ISBN: 9781942173762
Paperback | 6 x 9 | 256 pages | $22
Subjects: Ecology
⠀⠀⠀⠀⠀⠀⠀Politics
⠀⠀⠀⠀⠀⠀⠀Indigenous Studies

Frontline voices from the worldwide movement to decolonize climate change and revitalize a dying planet.

With a deep anticolonial and antiracist critique of what "conservation" currently is, Decolonize Conservation presents an alternative vision—one already working—of the most effective and just way to fight against biodiversity loss and climate change.

This powerful collection of voices takes us to the heart of the climate justice movement and the struggle for life and land across the globe. With Indigenous Peoples and their rights at its center, the book exposes the brutal and deadly realities of colonial and racist conservation for people around the world, while revealing the problems of current climate policy approaches that do nothing to tackle the real causes of environmental destruction.

THE RED DEAL
INDIGENOUS ACTION TO SAVE OUR EARTH

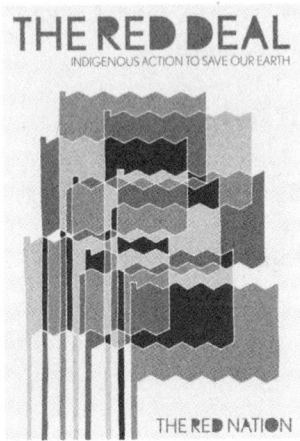

The Red Nation

ISBN: 9781942173434
Paperback | 5 x 7 | 176 pages | $15
Subjects: Indigenous Liberation
             Environmental Justice
             Social Movements

A powerful guide to Indigenous liberation and the fight to save the planet.

The Red Deal is a political program for the liberation that emerges from the oldest class struggle in the Americas—the fight by Native people to win sovereignty, autonomy, and dignity. As the Red Nation proclaims, it is time to reclaim the life and future that has been stolen, come together to confront climate disaster, and build a world where all life can thrive.

One-part visionary platform, one-part practical toolkit, The Red Deal is a call to action for everyone, including non-Indigenous comrades and relatives who live on Indigenous land.

Offering a profound vision for a decolonized society, The Red Deal is not simply a response to the Green New Deal, or a "bargain" with the elite and powerful. It is a deal with the humble people of the Earth; an affirmation that colonialism and capitalism must be overturned for this planet to be habitable for human and other-than-human relatives to live dignified lives; and a pact with movements for liberation, life, and land for a new world of peace and justice that must come from below and to the left.

BORDERTOWN CLASHES, RESOURCE WARS, AND CONTESTED TERRITORIES
THE FOUR CORNERS IN THE TURBULENT 1970S

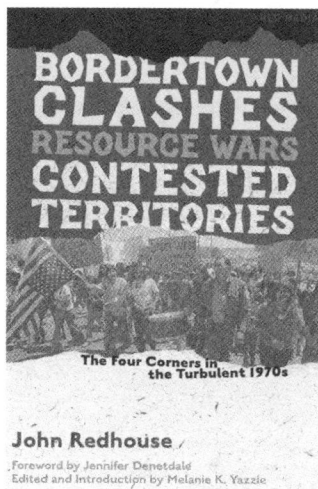

John Redhouse
Foreword by Jennifer Denetdale
Introduction by Melanie K. Yazzie

ISBN: 9781945335273
Paperback | 6 x 9 | 224 pages | $22
Subjects: Indigenous History
        Social Movements
        Ecology

A one-of-a-kind lyrical and fast-paced memoir of the frontlines and trenches of Native liberation in the Four Corners and Southwest in the 1970s.

From the late summer of 1972 to the late summer of 1974, Red Power activists put everything on the line to organize mass movements and direct actions for Native liberation. Written in the first-person with a spirit of generosity and witness, John Redhouse describes the fever pitch of the times, focusing on the racist and exploitative bordertowns in the Four Corners area of the Southwest region. Interweaving a piercing analysis of violence against Navajo people in reservations and bordertowns with a condemnation of the violence that rapidly growing mineral extraction in and around the Navajo Nation introduced to Navajo life.

As a firsthand participant in some of the most important twentieth-century struggles against this manifold violence, Redhouse is one of only a few grassroots intellectuals who can tell this story. *Bordertown Clashes, Resource Wars, and Contested Territories: The Four Corners in the Turbulent 1970s* brings readers to an extraordinary time defined by stunning victories and intense struggles that changed Native people's collective destinies.

ECHOES OF THE WAR WAR
LEGACIES OF COCHABAMBA BOLIVIA

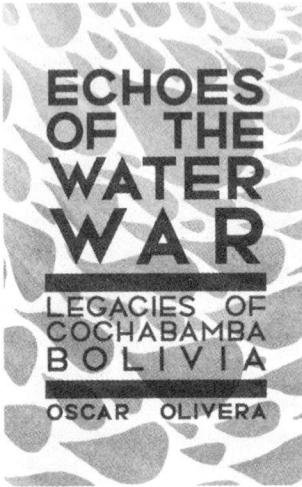

Ocar Olivera

ISBN: 9781945335297
Paperback | 5.5 x 8.5 | 208 pages | $18
Subjects: Latin America
          Social Movements
          Ecology

Lessons from one of the greatest people's victories against corporate neoliberal capture.

From the frontlines of one of the greatest popular rebellion against the privatization of natural resources comes the triumphant grassroots struggle that transformed ordinary people into water warriors. In a series of protracted struggles for direct democracy and defense of the commons between December 1999 and April 2000, the people of Cochobamba, Bolivia won popular control of water supply and defied all odds by driving out the transnational corporation that had stolen their water.

Oscar Olivera, a trade union machinist who helped shape and lead a movement that brought thousands of people to the streets, powerfully conveys the perspective of a committed participant in a victorious and inspirational rebellion. Alongside Olivera's hard-won political savvy, *Echoes of the Water War* presents leading voices on Latin American movements, including Raquel Gutiérrez Aguilar, Massimiliano Tomba, Raul Zibechi, Marcela Olivera, Stefano Archidiacono, Alexander Dwinell, and Nelly Perez. Together they reflect on major themes that emerged from the battle over water twenty-five years later.

IN DEFENSE OF COMMON LIFE
THE POLITICAL THOUGHT OF RAQUEL GUTIÉRREZ AGUILAR

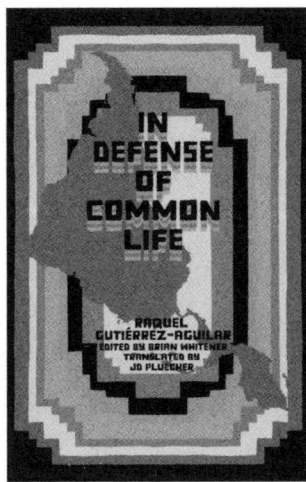

Raquel Gutiérrez Aguilar
Edited by Brian Whitener
Translated by JD Pluecker

ISBN: 9781945335112
Paperback | 5 x 7 | 176 pages | $18
Subjects: Latin America
        Indigenous History
        Feminism

"Raquel Gutiérrez Aguilar's brilliant analysis on the possibility of the commons helps us find ways to re-orient ourselves amidst the uncertainties of the political moment we are living in. Her nuanced and deeply critical political vision sheds light on what is enabled and possible when we look at the commons as the horizon for a politics centered around the very possibility of (re)production of life on earth. Critical of the different forms of complicity with neoliberal capitalism that leftist administrations have enabled, she looks at the complexity of articulating forms of collective desire that can channel social wealth to materialize ways of being in common.

Translating these works by one of the most important contemporary Latin American thinkers of the moment, allows us to understand different lines of history that permeate the question of the common, while facing the recurrence of state and capitalist captures of collective energies. Without posing easy formulas, Raquel offers critical trajectories to understand histories that are not much known in the North, thus open the possibility of inspiring readers to connect with and plot around key issues that traverses different geographies."—**Susana Draper**, author of *Libres y Sin Miedo: Horizontes feministas para construir otros sentidos de justicia* and coeditor of *Feminicide and Global Accumulation*

FOR ANTIFASCIST FUTURES
AGAINST THE VIOLENCE OF IMPERIAL CRISIS

Edited by Alyosha Goldstein
and Simón Ventura Trujillo

ISBN: 9781942173564
Paperback | 6 x 9 | 304 pages | $24
Subjects: Authoritarianism
          Imperialism
          Resistance

"This extraordinary volume ranges over a planetary geography and deeply engages historical formations and trajectories of fascism and antifascism. The authors, writing in a variety of genres and from many fields of study, illuminate the makings of racialized violence, the role of untruths, post-truths, and ideologies, the afterlives and ongoing effects of colonial force, and the role of capital accumulation in the making of modern varieties of fascism. Every page of *For Antifascist Futures* forces us to face and reckon with the lacerating effects of fascist power on the body politic"—**Laleh Khalili**, author of *Sinews of War and Trade* and *Time in the Shadows*

"Globalizing and reframing fascisms on a world scale, this urgent and powerful volume analyzes fascism as the convergence of authoritarian state and extralegal racial nationalist violence responding to the historical and material crises of capitalism and imperialism. The collection constellates a stunning range of antifascist practices, from Black radical internationalism, anticolonial movements, and insurgencies in the Philippines, Palestine, and South Asia, and across Latin America and Africa, on the one hand, to a long history of antifascisms and racial justice movements in the U.S. and Indigenous demands for return of stolen land, on the other."—**Lisa Lowe**, author of *The Intimacies of Four Continents*

HOPE AGAINST HOPE
WRITINGS ON ECOLOGICAL CRISIS

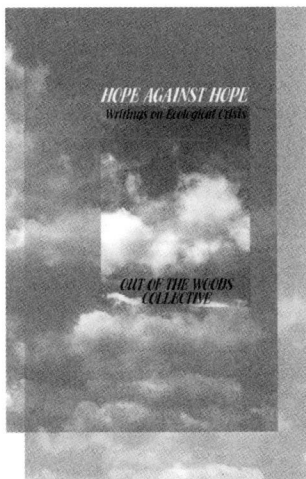

Out of the Woods Collective

ISBN: 9781942173205
Paperback | 6 x 9 | 272 pages | $20
Subjects: Climate
        Ecology
        Social Movements

Climate disaster is here. Capitalism can't fix it, not even with a Green New Deal. Our only hope against hope is disaster communism.

In *Hope Against Hope*, the Out of the Woods Collective investigates the critical relation between climate change and capitalism and calls for the expansion of our conceptual toolbox to organize within and against ecological crisis characterized by deepening inequality, rising far-right movements, and—relatedly—more frequent and devastating disasters. While much of environmentalist and leftist discourse in this political moment remain oriented toward horizons that repeat and renew racist, anti-migrant, nationalist, and capitalist assumptions, Out of the Woods charts a revolutionary course adequate to our times.

    At the center of the renewed political orientation Hope Against Hope expounds is an abolitionist approach to border imperialism, reactionary ecology, and state violence that underpins many green solutions and modes of understanding nature. Their stunning conclusion to the disarray of politics in our seemingly end times is the urgency of creating what Out of the Woods calls "disaster communism"—the collective power to transform our future political horizons from the ruins and establish a climate future based in common life.

BECOME A COMMON NOTIONS MONTHLY SUSTAINER

These are decisive times ripe with challenges and possibility, heartache, and beautiful inspiration. More than ever, we need timely reflections, clear critiques, and inspiring strategies that can help movements for social justice grow and transform society.

Help us amplify those words, deeds, and dreams that our liberation movements, and our worlds, so urgently need.

Movements are sustained by people like you, whose fugitive words, deeds, and dreams bend against the world of domination and exploitation.

For collective imagination, dedicated practices of love and study, and organized acts of freedom. By any media necessary. With your love and support.

Monthly sustainers start at $15 and receive each new book in our publishing program.

commonnotions.org/sustain